Mapping Israel, Mapping Palestine

INSIDE TECHNOLOGY

Inside Technology

edited by Wiebe E. Bijker, W. Bernard Carlson, and Trevor Pinch

A series list appears at the back of the book.

Mapping Israel, Mapping Palestine

How Occupied Landscapes Shape Scientific Knowledge

Jess Bier

The MIT Press
Cambridge, Massachusetts
London, England

This book was set in ITC Stone Sans Std and ITC Stone Serif Std by Toppan Best-set Premedia Limited. Printed and bound in the United States of America.

Library of Congress Cataloging-in-Publication Data

Names: Bier, Jess, 1980- author.
Title: Mapping Israel, mapping Palestine : how occupied landscapes shape
    scientific knowledge / Jess Bier.
Description: Cambridge, MA : The MIT Press, [2017] | Series: Inside
    technology | Includes bibliographical references and index.
Identifiers: LCCN 2016039117 | ISBN 9780262036153 (hardcover : alk. paper)
Subjects: LCSH: Cartography--Political aspects--Israel. |
    Cartography--Political aspects--West Bank. | Cartography--Political
    aspects--Gaza Strip. | Land settlement--West Bank--Maps. | National
    security--Israel--Maps. | Arab-Israeli conflict--Influence. | Map reading.
Classification: LCC GA1323.7.A1 B54 2017 | DDC 526.095694--dc23 LC record
available at https://lccn.loc.gov/2016039117

10 9 8 7 6 5 4 3 2 1

In memory of Oliver Milton Bier

Contents

Acknowledgments

I began this book by redefining my geographic boundaries to include Maastricht. My dissertation advisers, Sally Wyatt and Bas van Heur, were incredibly supportive from the start, and I am infinitely thankful for their critical guidance and friendship. They were joined by my former colleagues in the technology and society department at Maastricht University (MUSTS), many of whom read chapters from this book in draft form. Matthijs Kouw has provided invaluable encouragement, as have numerous other colleagues and friends who have greatly enriched my time, including José Cornips, Birgit Bertram, Lonneke Theelen, Gili Yaron, Alexandra Supper, Koen Beumer, Eefje Cleophas, Anna Harris, Thomas Fuller, Jessica Mesman, and Veronica Davidov. The Netherlands Graduate Research School of Science, Technology, and Modern Culture (WTMC) offered a welcome space for further discussion with colleagues and mentors, including Teun Zuiderent-Jerak, Willem Halffman, Huub Dijstelbloem, Geoffrey Bowker, Helen Verran, and so many others. Maastricht University and the WTMC also graciously supplied the funding for this research.

My fieldwork was made possible through the generosity of many in Palestine and Israel who shared their time and expertise. Jaad Isaac and the colleagues of the Applied Research Institute–Jerusalem (ARIJ) provided vital assistance and gracious hospitality. Riham Dweib answered endless questions and made me feel incredibly welcome. Ayman Abuzahra did so as well, and graciously has allowed me to use a photograph of him to illustrate the work done at ARIJ. Many were equally welcoming, including Ahmad, Issa, Sari, Ayman, Laila, Raed, and Suhail as well as Stefan Ziegler and the staff of the Border Monitoring Unit (BMU) at the United Nations Relief and Works Agency for Palestine Refugees in the Near East (UNRWA). Numerous cartographers and geographic information science (GIS) specialists explained their work to me with great patience. Some of them guided me back and forth across hills, checkpoints, walls, and innumerable borders—Riham in minibuses, Hagit Ofran in her truck, and Sahar and Rotem on foot, not to mention in taxis as well as on scooters and bicycles, but never a Segway. Lilach, Lika, Tammy, Taina, Shelby, Zuzanna, Fanny, Dan, Katie, and Blake sent regular invitations and welcomed me into their homes.

This book is much improved through the painstaking support of my editor, Katie Helke, and the editors of the Inside Technology series, Wiebe E. Bijker, W. Bernard Carlson, and Trevor Pinch; the thoughtful comments of the anonymous reviewers at MIT and beyond; and the efforts of my dissertation committee at Maastricht University. In my current position as an assistant professor of urban sociology at Erasmus University Rotterdam, I have the privilege of continuing to work on the key concerns that motivate this book, albeit on new topics. Willem Schinkel is truly inspiring both as a mentor and friend, as are Sanne Boersma, Rogier van Reekum, Maja Hertoghs, Irene van Oorschot, Friso van Houdt, Anne Slootweg, and Eva van Gemert. Bethany Hipple Walters, with the help of Carey, makes life in Rotterdam so much better with her kindness and friendship. Pamela Malpas provided irreplaceable and careful guidance that made the book possible in its present form. Miriyam Aouragh, Charles van den Heuvel, Thomas Pablo Sciarone, and Wytske Versteeg each contributed their thorough attention.

A number of scholars took time to talk and drink coffee or tea with me, even when I could offer nothing but questions and more questions, including Khalil Tufakji, Anat Leibler, Dov Gavish, Oren Yiftachel, Chris Parker, Omar Salamanca, Khaldun Bshara, Chen Misgav, Orit Halpern, Elena Glasberg, Ihab Saloul, Kristín Loftsdóttir, Irna van der Molen, Shelby Carpenter, and Sofia Stamatopoulou-Robbins. Christine Leuenberger assisted greatly with fieldwork and generously ensured that this research got off to a rapid start. Faculty members at Birzeit University and Tel Aviv University also spoke with me at length about their work. With Riham's help at Birzeit, they also saved a place for me in the bus for field workshops and graduate seminars. The organizers of many conferences offered valuable comments, including those of the Postcolonial Europe Network conference: Postcolonial Transitions in Europe, along with Ulrich Best, Noam Leshem, Alisdair Pinkerton, Noura Alkhalili, Salvatore Paolo de Rosa, and Anders Lund Hansen. Staff members at numerous archives were invaluable, helping in the name of research even when they were not entirely sure what the product would be. They include the directors of the ARIJ Library, Oded Fluss at the National Library of Israel, Dora Druskin and the employees of the Tel Aviv University Map Library, Hana Pinshow at the Ben Gurion Research Institute, and the members of the GIS Unit at the Israeli Central Bureau of Statistics (CBS).

These concerns have been with me for the better part of my life. Academically, this project began when I studied at the American University in Cairo in fall 2000. It took root during my time at the City University of New York Graduate Center, where I also worked as a freelance GIS specialist. Marianna Pavlovskaya was instrumental in my development as a writer, as were Cindi Katz, Joseph Massad, William Tutol, Elinore Pedro, and Sean Tanner. Many others were generous in ways that ultimately benefited this book whether they knew it or not. They include Gamil Youssef, Anne Meneley, Mark Monmonier, and Sarah Gualtieri. Neil Smith and Michael W. Suleiman passed away too soon, and they are greatly missed. Those named here may or may not endorse the final book or its arguments, but I am truly thankful for their help.

When I was a child, my parents never refused to buy me a book and often went to great lengths to find specific ones that I requested, with my mother combing the stacks at local libraries, and my father scouring bookstores during his travels. This book wouldn't exist without their tremendous support and encouragement. Ben and Ethan, Rachel, Lucas, Carter, Natalie, Declan, and Charlotte force me to put down my laptop now and then, and I wouldn't want it any other way. Our boisterous and ever-growing extended family in California and New Jersey made sure I got into enough trouble to keep things interesting, but never too much that it wasn't possible to get out of it again. I'm always really happy to see them.

Jayne DeBattista enriches every moment of my life. We met in the course of this research, and she serves as an ongoing reminder of what empathy and respect can be. Every experience is better in her orbit. Her impact goes so far beyond this book, and she read every word—many times over. Amy Hay Howard has long humored my love of obscure technical details, which she partly shares as a professional and aficionado of all things animated. Thanks to her I've gotten to know the wonderful Andy, Sasha, and Alice. So many other friends have contributed to the ongoing process of cooking, laughing, and bold imaginings, some of them going back to the days of the now-defunct *Tooth and Nail* magazine. They include Alejandra Gonzalez, Amira Mittermaier, (and beautiful Felix!), and Samantha Severin. Claude the cat always kept my writing warm by sleeping on top of the keyboard, whether I was using it or not. In fact, she's doing so as I type this sentence. Lastly, my fieldwork coincided with so much suffering, in Syria, Iraq, Afghanistan, Pakistan, Libya, Egypt, Palestine, Israel, France, Norway, the United States, and beyond. It included the devastating loss of my young nephew, Oliver, before his first birthday to a rare form of brain cancer. This book is dedicated to him. I am eternally grateful for the time we had together, even though it was far too short. More than ever, he reminds me of the importance of every single day and every single life.

A Note on Translation

All translations are my own, except where indicated. When transliterating Arabic and Hebrew terms, I used accepted English spellings for proper nouns, geographic features, titles of textual sources, and common words, in cases where they exist. I omit diacritics for the purpose of readability, given that most of the terms are in widespread use. In cases where particular individuals have provided their own preferred transliterations of their names, I use those. Otherwise I follow the American Library Association–Library of Congress system, with minor variations for readability depending on the context.

List of Abbreviations

ARIJ	Applied Research Institute–Jerusalem
BDS	Boycott, Divestment, and Sanctions Movement
CBS	(Israeli) Central Bureau of Statistics
GIS	geographic information science
GRPs	Graphical Rational Patterns
IDF	Israeli Defense Forces
LULC	land use / land cover
MOP (MOPIC)	PA Ministry of Planning, formerly the Ministry of Planning and International Cooperation
NGO	nongovernmental organization
OPT	Occupied Palestinian Territories
PA	Palestinian Authority
PACBI	Palestinian Campaign for the Academic and Cultural Boycott of Israel
PCBS	Palestinian Central Bureau of Statistics
PLO	Palestinian Liberation Organization
PPIB	Physical Planning and Institution Building
SOI	Survey of Israel
SOP	Survey of Palestine
STS	science and technology studies
UNOCHA	United Nations Office for the Coordination of Humanitarian Affairs
UNRWA	United Nations Relief and Works Agency for Palestine Refugees in the Near East

1   Where Cartographies Collide

Just as none of us is outside or beyond geography, none of us is completely free from the struggle over geography.

—Edward Said, *Culture and Imperialism*

On the Variegated Use of Maps

"We don't use maps in Palestine," a colleague told me soon after my return to the region in 2011 to conduct the fieldwork that would lead to this book. "It's a small country. We know where we're going." He was mostly joking, but it was a refrain that I was to hear many times during interviews. Maps make interesting decorations to put on a wall or useful tools for political negotiations. Maps of political borders are also routinely used as potent symbols for political movements—but they're not necessary for getting around. This attitude stems in part from the fact that maps of the region are often misleading. Although its location has since been updated, the Western Wall, one of the most famous religious sites in the world, was misplaced on Google Maps as recently as 2010. Instead of being located in the southeastern area of the Old City, it appeared almost one kilometer off target, in East Jerusalem (figure 1.1).

The process of using maps can be confusing due to the constant changes in the landscape. In the context of the ongoing Israeli occupation of East Jerusalem, the Golan Heights, the West Bank, and the Gaza Strip, Israeli military roadblocks and checkpoints are regularly enforced. The calculated unpredictability of these obstructions has made many maps, such as the UN closure maps (e.g., UNOCHA 2008), obsolete as travel tools before they are even printed. Where travel maps of the West Bank do exist, they are produced primarily for tourists and religious pilgrims from elsewhere. In an

Google maps

| western wall, jerusalem, israel | ▼ | Search Maps | Show search options |

Get Directions My Maps   «   🖶 Print ✉ Email 👁 Link

Western (Wailing) Wall
Western Wall Plaza, Jewish Quarter, Jerusalem, Israel
02-6271333
★★★★★ 11 reviews
Directions Search nearby more ▼
Category: Attraction

Pan⊕ramio

"Israel is AMAZiNG!" - tripadvisor.com ... "When we came back she stopped" - tripadvisor.com ... "It is a beautiful scene Friday at sundown" - tripadvisor.com ... "The men were joyous and dancing and hollering and having a great time" - tripadvisor.com ... "It kind of sounded eerie" - tripadvisor.com ... "The Wailing Wall is a MUST!" - tripadvisor.com ... "The wall is plain, and simple" - igougo.com

Sponsored Links

Tour Jewish EastJerusalem
Har Hazeitim, Shimon Hatzaddik, E1,
Shepherd Hotel, Demograhics History

Figure 1.1
A Google Maps Screenshot from June 29, 2010. This shows the Western Wall placed mistakenly by the Damascus Gate to the north, instead of by the Temple Mount in the Old City's southeast corner. The error has the effect of taking one of the most potently symbolic sites of Israeli worship and situating it in the heart of Palestinian East Jerusalem.

interview, the vice president of a local cartographic company noted that the "street map and the tourist map, they are both the same for us, because it's all for tourists." He stopped to add that local corporations occasionally do commission wall maps to give as gifts (interview 2, cartographer in private Palestinian geoinformatics firm). Those I spoke with emphasized the importance of interpersonal relationships, instead of maps, in finding one's way, both literally and metaphorically. Indeed, although people did have apps like Google Maps on their phones, many pointed out that the best way to get around was simply to stop and ask for directions.

By contrast, in Israel the use of maps is more common as part of everyday life, but it doesn't exclude other means of navigation. Even the spread of digital cartographic technology, intended in part to make cartography more accessible, has not precluded recourse to wayfinding methods that don't involve maps. On the English Web site for *Ha'aretz*, perhaps Israel's best-known newspaper, a commenter criticized the government's unsuccessful legal battle to prevent Google Street View from coming to Israel. Street View allows Internet users to look at photographs of streets that are taken at ground level. It's a way of getting a picture of a specific address. Its opponents worried that if Street View were available in Israel, then it would make it possible to find sensitive sites like the prime minister's mansion (Hasson 2012a; Yaron 2011, 2012, Zrahiya 2011). Yet the commenter responded that such locations could be found easily without a map: "Thousands of Israelis demonstrate at the [prime minister's] residence. Terrorists can just ask a cab driver or anyone in Jerusalem" (Hasson 2012a).

The comment is interesting for two reasons. First, it dispels some of the paranoia with respect to maps as secret repositories of classified information—an attitude that often contributes to maps' limited effectiveness in contested areas. Second, it appeals to observation, which is presented as common sense—something that is equally accessible to everyone in the landscape. Overall, it suggests that maps, however central to political debates, are not absolutely necessary for navigation. "Terrorists" can simply ask for directions to landmarks that "anyone" can see.

In this book, I question whether there are objects in the landscape that just anyone can see. I show how something that might be readily apparent to one observer might not be obvious to another, and seek to better understand how these differences have come to be. This is important far beyond the study of empirical maps, and far beyond maps more generally, because maps are often taken as stand-ins for science writ large. Scholars like Karl Polanyi and Thomas Kuhn draw analogies between maps and scientific theories (Livingstone 2003, 160). They do so because a belief in the triumph of one "true" observation, and one single best map, is frequently expressed together with a belief in technoscience as the best possible form of knowing. Therefore, to become immersed in maps is to study the very constitution, naturalization, and domination of specific kinds of knowledge over others. This is particularly evident in the case of digital maps. Digital maps are touted as the "end of history" when it comes to cartography. It is argued that they have the power to solve all the glitches and erasures of paper maps, to make the world known in all its raucous detail.

The idea that there could someday be one supremely accurate map is related to the notion that Earth is a singular space that is both definable and knowable. Astronauts are said to have formed the first comprehensive views of the planet as a single globe. This *overview effect* led them to see the irrelevance of national borders and look forward to one unified world. It is an evocative perspective that still resonates far and wide—one that has been built into the ways that the world has been mapped ever since. But it also papers over the significant role that nations played in making possible this penultimate picture of Earth, this idea of a universal map—a picture that is thoroughly grounded in the politics of the Cold War and idea of a global empire. Nation-states are precisely the backers of the organizations that trained the astronauts and funded their travel. Nations continue to shape so much of the cartographic data in use today, including those produced for private corporations. Their efforts include regulating the ownership and standard settings of data satellites—a process that requires a corporation like Google to operate in innumerable countries at once to amass their global data.

Furthermore, despite the proliferation of drones and satellites, geographic facts and statistics continue to be based on observations on the ground—a process geared toward interpreting and validating what is viewed from above in aerial photos (see chapter 5). Ground data collection is still central to digital cartography, and this largely takes place within the territories and legal frameworks of particular nation-states. Those nations are an excellent example of how historical and geographical entities can shape the very ways that people see, even making it possible to see one specific vision—say, of a 1960s' astronaut peering through a porthole as dawn rises over the Mediterranean—as the ultimate vision, as something global and universal. In contrast to this type of singular vision, in this book I argue that no level of accuracy, resolution, or detail will ever be sufficient to form one overall, authoritative picture of the globe. More and better data will never make either maps or theory behave, nor fully tame their troublesome behavior. But this trouble can be seen in a positive light, because it means that the world is a far more interesting place than it might otherwise be. It also means that people will forever be devising new ways to both think about the world and live in it.

An International Occupation

To get at the varieties of vision across nations, this book maintains an international frame. It considers both Palestinian and Israeli accounts as well as ones that don't fit neatly into a divide between Palestinians and Israelis. Of course, this international frame does not escape nationalism or state institutions. Quite to the contrary, it allows for an investigation into the boundaries and rhetoric of individual nations and territories in a broader region where there are ongoing military interventions, including the Israeli military occupation of the West Bank and Gaza Strip, among other territories.

An international or supranational framing is crucial in situations where sovereignty is under constant challenge, and for groups like the Palestinians that have yet to be granted a fully independent nation of their own. Thus *internationalism*—specifically, the forms of control inherent in international knowledge and their relationship with colonial and military legacies—is the first of three core themes of this book. After presenting it here, I move on to the second theme of *landscape*, followed by the third and final theme of *symmetry*, which is used precisely to bring stark asymmetries of power into relief. I then conclude with some broader background of the region as told through the lives of two pivotal twentieth-century cartographers.

The chapters that follow provide a combined analysis of two central types of international landscapes. First, there are the transnational landscapes of science and technology, or *technoscience*—including related practices that might alternately support or seek to end continued injustice. Second, there are the occupied landscapes of Palestine and Israel. The use of *international* here doesn't simply refer to everything beyond a single

nation-state, however. It instead points to specific forms of internationalism as well as the particular relations among nation-states that developed
during the nineteenth and twentieth centuries, and are epitomized by
supranational organizations like the United Nations. The United Nations
and its ancestors have played decisive roles in the current political status of
Palestine and Israel. In UN debates, it is often suggested that rational governance and well-defined cartographic boundaries are the most effective ways
to resolve the Israeli–Palestinian conflict. Yet I demonstrate that the effects
of this call to rationality are rarely, if ever, so simple.

The maps are not simple either—but there are plenty of them. The local
non-use (Wyatt 2003, 2008a) of maps doesn't mean that there are no maps
of Palestine and Israel. Just the opposite: Jerusalem has long been one of
the most regularly mapped places on Earth. The earliest maps date at least
from the Madaba mosaic in Jordan in the sixth century CE—making the
region one key fulcrum in global histories of science. The maps, though,
are frequently both highly imaginative and highly selective in terms of the
features and areas that they display. In recent decades, they often have
been created as part of the struggle over facts in political disputes. In addition, maps play an important role in swaying the broader public perception of the conflict. This again highlights the international dimension of
cartography in a region that is the focus of the work of numerous international technicians, political organizers, aid workers, medical professionals, reporters, well-known cartographers like Jan de Jong, and many others.
The politics of Palestine and Israel are similarly internationalized, from
solidarity activists who work with international Palestinian organizations,
to Jewish groups abroad with ties to Israel, to those who are employed
by organizations like the European Union. Internationals have long histories of involvement there—involvement that stretches back even further
if we consider the area's millennia-long history as a site of religious pil-
grimage from around the world. All told, the region is tremendously
important internationally, especially for an area that is roughly the size of
the US state of Vermont (Hadawi 1957) or about half the area of the Netherlands. This book centers on Palestinian and Israeli cartographers because
they are typically the ones with the longest histories of making maps of
Jerusalem and its surroundings. But as technical elites, many of whom also
have roots in diaspora, their lives and cultures generally are international
as well.

This book also focuses on *empirical* maps, or maps that are based on observations and sensory experience. Such maps are intended to show the landscape as it currently *is*, in contrast to those who aim to make subjective maps, including artistic maps that aim to depict it as the cartographer wishes it *could* or *should* be. This division between empirical and subjective maps is far from definitive, but it reflects a distinction that is present within the field of cartography itself. Many of those I spoke with stressed that they were presenting a scientifically accurate view of the landscape and were creating maps that displayed, in their view, objective evidence for the region (such as population statistics). This is crucial because no one expects subjective maps to all agree, since they reflect different opinions and imaginations, but it is surprising to many when supposedly objective, empirical maps differ widely from each other. One central aim of this book, therefore, is to interrogate the kinds of disagreements that occur even among cartographers who believe that their work is based on objective observations that, to their minds, anyone might see to be true.

These empirical practices are linked to a particular form of scientific internationalism. Indeed, the introduction of empirical computer mapmaking has reshaped the international bent of many maps in the region. Digital cartographic technology, such as GIS mapmaking software, is dominated by corporations that originate in Europe and North America, and the use of GIS has helped to tighten the hold of international science on cartography in Palestine and Israel.[1] Yet this is far from a simple one-way influence or transfer of technology from the West to the rest of the world. Early in its own development, digital cartography was brought to Palestine and Israel as part of the trend toward computerization in the mid- to late twentieth century. It was intended to help broaden the pool of cartographers, in part by reducing the initial costs necessary to begin making maps. As such, GIS was also developed in Palestine and Israel from an early date, not least as a military technology. While nonetheless still expensive, its use—in combination with the increasing availability of free, online mapmaking tools—has led to the further incorporation of Palestinians and Israelis from diverse backgrounds. In the process, GIS has helped to found new digital paradigms for traveling through space and viewing the world.

Early twenty-first-century cartography includes a range of empirical practices in Jerusalem and the West Bank, from decommissioned spy satellite images to a road map made by Palestinian students who tracked their own movements on their mobile phones (Engineers without Borders Palestine, n.d.). But as GIS and cartographic practices have changed through their incorporation into the region, they have done so in part through the heavily segregated geographies of the Israeli occupation—geographies that the cartographers differently move through and inhabit. Physical boundaries reflect and reverberate across political divisions, shaping and separating communities of cartographers, thereby allowing for the creation of divergent methodologies and standards of practice. While there is no reason to believe that this might be more or less effective in Palestine and Israel than anywhere else, the specific influences of the regional geographies of Palestine and Israel highlight central issues in the use of GIS that are of interest everywhere that digital technology is used.[2]

How have such geographies shaped empirical cartographic practice across space and time? The way that knowledge is situated in social and material landscapes is the subject of this book. Through three representative cases on population, governance, and urban maps, I analyze how specific aspects of these segregated landscapes have affected the production of geographic knowledge of Jerusalem and the West Bank from 1967 to the present.[3] While technology is often believed to provide an objective perspective, this book contends that all knowledge is embedded in particular times, places, and cultures—including the international cultures of technoscience. Therefore, I critique the notion that technology functions as an impartial arbiter in disputes that hinge on empirical facts. Although ideology, nationalism, and religion are overwhelmingly important aspects of such disputes, the focus here is on knowledge defined as *empirical* and *scientific* precisely because it is widely believed to be objective and value free. Drawing on work in science and technology studies (STS) and geography, I show how even the most basic observations of the location of a border, or the height of a building, are also inherently subjective and political. This doesn't make those observations any less scientific. Instead, it calls for a different view, indeed a situated view (Haraway 1988), of knowledge.

The Significance of Segregated Landscapes

Yet that situated view is itself embedded in a variety of landscapes that are simultaneously social and material. So there is no escaping the influence of landscape, and landscape is the second core theme of this book. The notion of landscape emphasizes how important geographic space is in shaping the ways that knowledge is made. Examples of landscapes include border zones; urban streets and architecture; rural rain collection cisterns and land management systems; formal centers of knowledge like libraries, archives and government institutes, and community organizations; and places of confinement like checkpoints, prisons, and detention centers. Rather than focusing on practices within a predetermined geographic frame, such as a neighborhood, city, or nation, the study of geographic landscapes serves to highlight how geographic categories like "city" or "nation" are produced in discourse and practice. If knowledge is made in and of a particular place and time, those places and times are never isolated. Landscape, as I use it, is thereby an inherently relational and performative concept, meaning that it allows for research on how social and material practices pragmatically constitute reality in relation to one another across space, time, and forms of social difference. It enables the study of how boundaries and borders come to be. It also enables an analysis of the effects that borders have as they are socially and materially inscribed into the ground. As such, landscape stresses an understanding of the role of infrastructures (Aouragh 2011b; Salamanca 2014)—like roads, buildings, walls, checkpoints, and Internet servers—in shaping the ways that people move and the kinds of maps they make.

Together with an analysis of borders, the study of landscapes allows for an investigation of the role of geographic *scale*. Two conceptions of scale run throughout this book. First, I follow the late Neil Smith's (1992) call to avoid taking for granted the existence of particular scales, such as the national, global, or local. Instead, Smith urged scholars to recount how specific scales were produced as the natural or given levels for the study of society—a process that has been further explored by many, notably including Sallie Marston (2000), Anssi Paasi (2004), and Erik Swyngedouw (2004).[4] This is precisely the aim of examining landscapes that may or may not fall within the territory of one or more nations. Doing so acknowledges that scale is never fixed or stable. In this vein, Orit Halpern (2015, 35), drawing on the work of El Hadi Jazairy, calls attention to how scale is "plastic" and unstable, "a matter of ongoing relations" and "relationships between surfaces ... that are not automatically commensurate." Second, I build on the writings of recent scholars who critique the binary between local and global scales. The local and global are regularly used as contrasts to one another, yet often they coexist. Many kinds of knowledge and practice are local, national, and global—and so on—all at once. Similarly, there is no reason to only identify community groups with the local, and states and corporations with the global. Community practices may focus on local impacts while at the same time creating a global network of related organizations. Multinational corporations may span the globe yet consist of particular projects in small but numerous locales (Gibson-Graham 2002; see also Herod and Wright 2002). This simultaneity of many scales in turn frames my analysis of how international systems of power, which may have diffuse spatialities, might coexist with colonial methods that concentrate on defining and gaining control over specific geographic areas.

Asymmetrical Injustice and Occupied Knowledge

In addition to internationalism and landscape, *symmetry* is the third key theme of this book, as mentioned above. For reasons that are explored more fully in chapter 2, this book is structured symmetrically between (self-defined) Palestinian and Israeli accounts in an attempt to understand how knowledge itself becomes infused with the social and material effects of the Israeli occupation. As a symmetrical account, it involves setting Palestinian and Israeli maps and mapping practices side by side in order to then investigate which, if any, differences emerge. Yet this is not done in order to reinforce the notion that there is necessarily something naturally or essentially "Palestinian" or "Israeli" about a particular approach, or to suggest the two groups are mutually exclusive. Neither is it intended to imply that the best way to approach the conflict is to treat Palestinians and Israelis as equal but opposing partners in a verbal disagreement. Instead, symmetry throws into relief the systematic imbalances of power among Palestinian and Israeli cartographers—imbalances that hold even though both groups come from international technical elites.

Indeed, the spread of computer mapmaking has not, or not only, led to the democratization of knowledge in the region. Palestinian forms of empirical knowledge continue to be discounted in a number of ways. These include challenges to Palestinians' personal and professional identities, the hampering of Palestinian institutions and knowledge infrastructures, and the refusal to acknowledge Palestinian maps even when they fully comply with accepted international standards. So rather than suggesting that Palestinians and Israelis are treated equally, either locally or internationally, I employ symmetry precisely to better understand knowledge forms produced under occupation—forms that are relevant wherever military force bolsters scientific inquiry. Instead of excising the occupation from the study of science and technology in the region, symmetry explicitly highlights the unjust treatment that many suffer under the Israeli occupation. In keeping with the wider effort to achieve symmetry in STS (Wyatt 2008b; see also chapter 2), I temporarily and strategically set the categories *Israeli* and *Palestinian* equivalent to each other in order to show how unequally and unequivalently they are treated in practice. I thus join numerous antioccupation voices in calling for further reflection on the idea of an international dialogue among equals as the primary means of working to end injustice, given that often one of the parties to the dialogue is in a role of enforced subordination that gives the lie to notions of free debate.

For these reasons, the use of symmetry in this book differs from the kind of symmetry described by the Palestinian Campaign for the Academic and Cultural Boycott of Israel (PACBI). In their call for an international academic boycott of Israel, the PACBI (2009, 2011) states that "supporters are asked to refrain from participating in any event that morally or politically equates the oppressor and oppressed, and presents the relationship between Palestinians and Israelis as symmetrical."[5] Here the PACBI informatively notes that implicit symmetry is often used to suggest that the occupation is not a political struggle, one that has concrete causes and related power imbalances, but rather simply a disagreement over terms or definitions of the kind noted above. That narrow, and sadly more common, use of symmetry indeed papers over the process by which the participants who are coerced into cooperating in such discussions, the vast majority of them Palestinians, continue to be systematically stripped of their homes, lands, and lives. In contrast to but with great respect for their definition of symmetry, I use symmetry as a means to focus within and across the production of boundaries. This allows for an analysis of how group injustices sometimes can be reinforced through, not overcome by, technoscientific practice and knowledge more broadly. It also shows how such imbalances reverberate across international hierarchies that differently affect Palestinians and Israelis, and therefore serve to reproduce systemic injustices of the kind that the PACBI productively and effectively contests.

Accounts that represent the Palestinian–Israeli conflict as being the result of what Edward Said (1995, 36) called a "perhaps tragic, but certainly rectifiable, psychological misunderstanding" do little to address the ongoing material injustices of the Israeli occupation. To counter such facile attempts at commensuration, this book helps to further support Thomas Abowd's (2007, 243) point that although there is a "parity of desire" for Jerusalem and related areas among Palestinians and Israelis, so far "there has been no parity of power." In so doing, I focus on the work of professional elites precisely to analyze specific instances where Palestinians and Israelis, respectively, have actively participated in scientific debates. This does not mean that creating formal Palestinian or Israeli disciplines is the only way toward social justice. It nonetheless shows how, even when researchers fulfill the requirements of an international scientific or technological discipline, discrepancies between Palestinian and Israeli maps are unfairly interpreted as a personal failure or lack of will on the part of the Palestinians, and the Palestinians alone. Yet as I show, this routine characterization is unfitting on further examination. To this end, symmetry is a uniquely effective method of drawing attention to such imbalances, which persist even in such a "best-case" scientific scenario, where both parties aim to operate professionally, as recognized participants in an open, factual debate.

The technique of symmetry, as it is used here, enables an additional aspect of the analysis: namely, the investigation of the exclusionary effects of landscapes that serve to excise identities such as *Palestinian* and *Israeli* from one another, allowing for little overlap or in-betweenness (see chapter 5). For the categories of *Palestinian* and *Israeli* are also socially produced, although they have incredibly long and meaningful histories. Yet the harsh segregation of the occupation—not to mention the forced but unequal integration of the settlements—serves to steamroll over differences within groups, while polarizing differences between them. This combined process of polarizing and steamrolling discounts the experiences of those who do not neatly fit into any one group.[6] So although symmetry here examines what happens when scientists explicitly identify as *Palestinian* scientists, for example, or *Israeli* cartographers, it does not assume that these are, or must be, the only salient identities.

The Geographic Production of Knowledge or
Computers as Fact-Making Machines

These three themes of internationalism, landscape, and symmetry as well as the related notion of scale are further explored in chapter 2. Together they provide a unique approach to studying what I call the *geographic production of knowledge*, or the ways that knowledge is made in space and time. Such an analysis is crucial because contrary to the popular view, there is nothing natural or self-evident about "the facts." The act of creating a fact requires a huge effort of measurement, calculation, and definition. It involves making contact with and managing the social relations of factual institutions, like companies that sell GPS devices or conduct surveys, for instance. Many people take facts for granted, but facts never simply exist. Instead, they must be *made* (see chapter 5). STS scholars speak of the social construction of knowledge, and I combine this insight with literature in geography to shift toward the geographic production of knowledge. The notion of production allows for a focus, pioneered in STS, on the materiality of the practices of knowledge production as well as the incredible amount of labor and effort required to know any particular fact. It enables an exploration of the role of computers and digital technology in constituting facts. Yet the spatial character of these processes requires an understanding of not only the social and political contestation of different groups but also the ways that contestation takes place within and re-creates specific landscapes. My use of the term *production* therefore draws on Judith Butler's (2015) work on performativity and Annemarie Mol's (2002) account of enactment in medical practice. For this geographical study, however, I focus on production in order to highlight the material and temporal obduracies that are not reducible to speech, texts, or acts. Nevertheless, in keeping with conceptions of performativity and enactment, my analysis of production processes emphasizes how landscapes are never simply "out there." They must ever be remade anew (see chapter 2).

Maps relate to a broad swath of digital media and knowledge. For many people, daily life is awash with digital maps, and empirical maps are central to debates over the future of technology and society. Maps also are iconic technologies in Palestine and Israel, a central region in the history of map-making, whose role has only intensified since the advent of digital cartography has led to increasingly minute forms of surveillance and control (Abujidi 2011; Weizman 2007). Empirical maps and boundaries inform broader insights into the spatiality of all knowledge. They draw out what is contingent, amazing, and unexpected in knowledge-making practices that are too often taken to be natural and predetermined. Summarizing critical boundary theory, Janice E. Thomson (1996, 13) notes that "there is a marked tendency to accept boundaries as given, permanent, and even natural. ... On the contrary, they are arbitrary, contested, and ever-changing." This is also relevant to the borders of and within maps as one form of knowledge.

While Thomson is referring to the implementation of boundaries on the ground, a similar argument could be made for the boundaries displayed on maps, which are influenced by the cartographers' own relationships to the contested borders they draw. Intended to display objective facts, empirical maps frequently inspire tense discussions, with participants exhibiting a variety of observational frames that cannot be divorced from their unequal positions within the very terrains that they seek to portray. However, the purpose in this book is not to determine which locations of a border are true, or to expose the false claims of cartographers. Instead, it is to consider how knowledge is materially situated in ways that go beyond a dichotomy between true and false (Harley 2001; Swedenburg 1995). Above all, I seek to better understand the *how* questions: How are maps made, and how did they come to be made in these ways? This also touches on considerations of *who* and which groups are believed to be legitimate makers of knowledge, *what* or which kinds of knowledges—like facts, reports, and maps—are believed to be true or objective, *where* in the landscape are supposedly universal forms of knowledge made, and *why* knowledge is situated in these ways.

By elaborating on how knowledge is geographically and socially produced in landscapes, this book contributes to fields like STS, geography, sociology, anthropology, and political theory. Moreover, through an exploration of how borders have shaped the mobility of cartographers in a region that has become an "iconic model of geopolitical spatial dilemmas" (Abujidi 2011, 313), this book is relevant to a growing body of literature on the roles of technology and academic knowledge in entrenching forms of control. The literatures include sources on the Israeli occupation (Abu El-Haj 2001; Aouragh 2011b; Leibler and Breslau 2005; Ophir, Givoni, and Hanafi 2009; Weizman 2007; Zureik, Lyon, and Abu-Laban 2011) as well as a broad variety of studies of the intersections of power and expertise (Foucault 1986; Haraway 1988; T. Mitchell 2002).

To begin to see how earnest efforts to rationalize and simplify a complex political context might instead only lead to further complexity, it is necessary to investigate the impact of landscapes that have differently affected the lives of cartographers and their maps. With this in mind, I next provide a brief account of the history of cartography in Palestine and Israel through the lives of the cartographers Sami Hadawi and David Amiran. This is followed by a discussion of the method of *traveling ethnography* that I developed in the course of writing this book, but that nonetheless builds on prior ethnographic work.

A Nation of Scattered Documents: Colonialism and Cartographic Knowledge

Empirical cartography in Palestine and Israel is thoroughly international, and this is in no small part because the landscapes of the region are thoroughly colonial. Hadawi and Amiran were two cartographers whose experiences illustrate how colonial legacies can inform complex systems of segregation and injustice—and through them, the ways that knowledge is made. They were rough contemporaries, with their lives each spanning nearly the whole of the twentieth century, but changing imbalances of power shaped them in distinctive ways. Hadawi, a Palestinian scholar, spent his time documenting the land and people of Jerusalem and the surrounding regions, and his work would later be taken up by famous Palestinian cartographers like Salman Abu Sitta (2004, 2007). Amiran, a Jewish academic who fled the rise of the Nazis in Germany, would go on to become one of the preeminent Israeli geographers of the postwar period, noted for his wide-ranging travels in the region.

Although both Hadawi and Amiran were refugees, Amiran's local journeys were only made possible by the exclusion of Palestinians from the areas that became Israel, including Hadawi and his wife. Bringing Hadawi and Amiran together permits an exploration of how unequal landscapes differently shaped their lives. It also makes it possible, in turn, to better understand how those local, national, and international landscapes were themselves produced as imbalanced systems of power (Bank and Van Heur 2007; Smith 1992)—albeit without falling into undifferentiated "macroconcerns" (Latour and Woolgar 1986, 17). The stories of Hadawi and Amiran thereby provide a fitting introduction to the final portions of this chapter. However, before turning to them, it is first necessary to trace the colonial legacies that so affected their work.

The long history of colonial cartography in Palestine and Israel provides a rich basis for comparison between historical and contemporary practices. To give just a few examples, in addition to a host of biblically inspired maps of the Holy Land as well as maps drawn by travelers to the region in recent centuries, several major cartographic projects were undertaken in the region beginning in the late eighteenth century. After the topographic maps made by the French military during Napoléon I's 1799 invasion, the first large-scale maps were produced under the Ottoman authorities in 1887 in connection with public development projects in the Hula Valley (Gavish and Ben-Porath 2003). In the same decade, the British-led Palestine Exploration Fund produced detailed maps to aid pilgrims and authorities (Levin 2006), paving the way for the British to assume colonial control in Palestine. Their expeditions were extended with systematic landownership surveys of the territory covered by the colonial British Mandate (1920–1948), the last major cartographic project in the region before the establishment of the state of Israel (Gavish 2005).[7]

The end of British rule was another new beginning for colonial knowledge. Israel's declaration of statehood took place on May 14, 1948, sparking a war and, ultimately, a forced Palestinian exodus from conflict areas that is known in Arabic as the *Nakba*, or catastrophe. Even while the war was ongoing, cartographers set to work developing Israeli state cartography. Barely one year later, David Ben-Gurion, the first prime minister, appointed a Governmental Names Committee, including key members of the preexisting Jewish Palestine Exploration Society, to determine Hebrew names for all the villages, rivers, lakes, roads, mountains, and other features. This initiative was part of an overall process whereby Arabic and Arabized names were actively erased because although they were thousands of years old in some cases, they were defined as foreign in the context of the Bible's description of the ancient geography of the region. Almost overnight, the toponymy was transformed, with the effect that by the 1960s, a standardized national geography of Israel could be said to have emerged through close collaboration among diverse groups of public and private sector cartographers (Abu El-Haj 2002; Srebro 2009). This simultaneously led to an overpowering sense of loss even for those Palestinians who were able to remain on the land, due to its sudden unfamiliarity (Scham 2003, 75).[8]

Figure 1.2
A base map showing Israel and the Palestinian Territories, including the West Bank and Gaza Strip. The map is drawn from UNOCHA (2010b) (map), and it is the responsibility of the author. For updated UN maps of Palestine and Israel, see http://www.ochaopt.org/mapstopic.aspx?id=20&page=1 (accessed April 2, 2016).

In 1967, Israel's victory during the Six-Day War drastically expanded the territory under Israeli control.[9] The Israeli military captured the Sinai Peninsula and large portions of the previously Syrian-controlled Golan Heights. In addition, it occupied East Jerusalem, the West Bank, and the Gaza Strip. Since the armistice at the end of the 1948–1949 war, the former two areas had been under Jordanian rule, and the Gaza Strip was under Egyptian administration. The West Bank, the Gaza Strip, and East Jerusalem would come to be viewed internationally as the potential core of a future Palestinian state, and they became the focus of political negotiations and humanitarian aid to the Palestinians.

After 1967, the Israeli government argued that all the conquered territories were part of Israel, and no longer showed their borders on Israeli maps.

The 1967 war thus did more than reinscribe political realities; it brought into question the Israeli state borders themselves. From being an avowedly homogeneous state—albeit one that was made homogeneous through the exclusion of Palestinian refugees—and a state whose contested areas were located outside Israel, after 1967 Israel began to officially "administer" the Palestinian Territories as part of its military occupation of them.[10] Unlike East Jerusalem and the Golan Heights, however, the Gaza Strip and West Bank have not been formally annexed to Israel. In this respect, if not in others, the Gaza Strip and West Bank are liminal spaces: they are under the control of the state of Israel to a greater or lesser extent, but they are not formally part of the state. As such, they have neither the permanent control nor political rights that formal annexation—in theory, if not in practice—would imply.

Between 1967 and the early 2000s, the Israeli "Civil Administration"— an English translation of the Israeli government's Orwellian term for the occupation's military authorities—took an active role in the daily management of most of the Gaza Strip and West Bank. Yet after the Oslo Accords in the mid-1990s, the newly created Palestinian Authority (PA) was granted jurisdiction over large cities and population areas within the Palestinian Territories. In addition, in 2004, Israel "disengaged" from Gaza. The government formally removed Israeli settlers from Gaza, but nonetheless the state continues to maintain nearly full control over entry into and exit out of the area. This has led Noam Chomsky (2012) to call Gaza the "world's largest open-air prison." Gaza also has been racked by a series of wars, attacks, and bombing campaigns, officially in retaliation for sporadic rockets shot into Israel from the Gaza Strip. The enforced and wholesale isolation of Gaza contrasts with the treatment of the West Bank, where there has been a proliferation of territories with overlapping bodies of law and varying degrees of official control (see chapter 4). Thus multiple, ambiguous forms of sovereignty have emerged. Yet they all take place within territories that were ostensibly part of the Israeli state—and this despite the official illegality of the Israeli occupation under international law (Weizman 2007).

Throughout the post-Oslo period, the mapping of the densely fragmented territories of the West Bank would become a central focus of cartographers working in Palestine and Israel. The complexity of changes in the landscape—a proliferation of borders of multiple sovereignties in relation to the sometimes ambiguous borders of the nation of Israel—evolved in concert with emerging GIS mapmaking practices that allowed for increasingly detailed depictions of ever more minuscule parcels of land. At least since the late 1970s, Palestine and Israel have been considered a key region for the emergence of high-tech industries, including geospatial technologies such as remote sensing and computer cartography. Far from being part of a unidirectional technology transfer from the West to the Middle East, Palestinian and Israeli cartographic practices made early contributions to the broader transition from handmade paper maps to the data-rich maps now in wide use through Internet sites such as Open Street Map and Google Maps. Digital mapmaking in Palestine and Israel therefore can be seen as both a response to the need to administer contested lands and an enabling mechanism that allowed for maps to be made at much greater resolution at ever-finer scales. This provides for increasingly minute maneuvering in political definitions of land (Weizman 2007).[11]

Although GIS often has been employed as a conflict resolution tool, the increased level of detail and alleged accuracy has led neither to a lessening of the importance of territory (Newman 2001) nor to a straightforward resolution of political debates. Despite the advent of computer cartography, cartographers regularly make competing factual claims about Palestine and Israel, and as such, they are a key region for the study of how scientific knowledge is produced. This is particularly apparent in light of the fact that GIS and digital cartography were developed in the context of the dawn of computers in the late twentieth century, and form an important part of the history of computers and media. Computers allegedly allow any location to connect with another, enabling anywhere to be mapped from anywhere, supposedly without the need to visit the locations being mapped. Rather than emphasizing the methodological break initiated through the development of computers, however, it is necessary to investigate how computer cartography also connects with prior practices, as is discussed in chapters 2 and 4, as this is key to understanding why many of the utopian promises of the early spread of computers and the internet, including the free flow of information and a purported end to political disagreements, have yet to be realized.[12] In part to show how the onset of computers shaped cartography, I now return to the two cartographers who would help to lay the groundwork for the later incorporation of GIS technology. Although their stories are by no means wholly commensurable, nonetheless they allow for an exploration of the combined material and social conditions of their lives. These include the landscapes *where* both they and their maps were allowed to live and travel.

Traveling Cartographer, Gathering Maps

The son of a Palestinian mother and Iraqi father, Hadawi was born in Jerusalem in 1904 and spent part of his childhood in Jordan after his father was killed fighting for the Ottomans in World War I. He was employed as a translator before returning to Mandate Palestine to serve under the British, who were conducting a survey in order to document claims to landownership in the context of disputes between Palestinians and members of the Jewish community. In 1948, Hadawi became a refugee together with his wife. They lost the home they had built in the Jerusalem suburb of Katamon, now part of West Jerusalem, along with most of their belongings. From 1948 until his death in 2004, Hadawi worked tirelessly to produce facts about the people of Palestine, this time focusing on property that had been confiscated by the newly created state of Israel.[13] So perhaps it is not surprising that Hadawi ended up in New York as the land specialist for the United Nations Conciliation Commission for Palestine in 1952.

At the United Nations, Hadawi attempted to combine all the diverse property records relating to Palestinian landownership in one place, in the hope that compensation one day might be granted. Much of his time was spent collecting and shipping key documents across borders with the goal of finding them a stable, accessible home. It was an attempt to change the international system from within by appealing to the same international laws and decrees that had overseen first British colonization, and then the Palestinians' forced dispersal. In the words of Izzat Tannous, the director of the Palestine Arab Refugee Office in New York City, where Hadawi (1957, n.p., emphasis added) was later employed, their shared goal was to "place before the American people, the Government [*sic*], and the United Nations, the *true* facts about the situation in Palestine."[14]

Out of necessity, then, Hadawi became a collector of documents, sending representatives to Jerusalem from the supposedly neutral territory of New York, fighting for the right to copy or obtain records that had been seized by the Israelis. In so doing, he struggled against inequalities in access. The Israeli state was given copies of many British records that had not been in its possession in the immediate aftermath of 1948 (Gavish 2005), yet the same achievement took over twenty years of consistent effort by individuals like Hadawi and representatives of governments in the Arab world. Documents were even the reason why Hadawi ended up a refugee rather than an Israeli citizen. At the time of the cease-fire, he was in East Jerusalem delivering a group of Palestinian land records to the Arab High Committee, the main political body on the Arab side. If instead he had been able to stay in the general proximity of West Jerusalem, where his home was, and had transferred the documents to a central depot as per British orders, then Hadawi might have been made an Israeli citizen after the cease-fire, and the documents would most likely have ended up in the hands of the Zionist forces (Fischbach 2003, 213–214).

Building the Lab, Clearing the Field

Amiran was born in Berlin in 1910, making him just six years younger than Hadawi. He too became a refugee when after leaving Germany to work as part of a scientific survey in Moscow, the rise of the Nazis made it difficult for him, as a Jewish person, to return. He moved first to Switzerland to complete his PhD and then to Israel—thereby avoiding being in Germany during the Holocaust. Amiran arrived in Jerusalem in 1935, and lived the rest of his life in the city and its environs. He enlisted in the Mapping Unit of the British Army during the Second World War, serving the Allied effort in Egypt. He then served in the Israeli Defense Forces (IDF) in 1948, becoming the deputy commander of the Mapping and Aerial Photos Unit before helping to found the Department of Geography at the Hebrew University in Jerusalem (Bar-Gal 2013). In this capacity he served as a bridge between the military, academia, and the state. He simultaneously furthered his scientific research and institution-building efforts. Amiran legally may have had the ability to return to live in Germany after World War II and the devastation of the Holocaust, but he never did so, perhaps also because he was faced with the violent erasure of Jewish people, towns, and cultural practices since the Jewish communities in Europe had been so aggressively decimated.

After 1948, Amiran worked to construct Israeli academia as a national system that snugly fit into and yet struggled to participate as an equal member within international scientific communities. In the process, he became a noted expert on arid regions, arguing against those who claimed that Israel was a desert and therefore unfit for modern development.

Amiran (1987) sought to prove them wrong, discursively by contending that Israel was a desert only as a result of past Arab mismanagement and materially by helping to redevelop Israel's landscapes. Both projects required extensive travel within Europe, North America, and even the USSR in order to obtain support and funding for his research. Amiran's trips in turn were financed and legitimated by the economic reshaping of Israel along lines that fit Western conceptions of modern development, as I will discuss in chapter 2.

Amiran also traveled locally to collect data and measurements of the natural landscape. However, his local trips were made possible by the measures taken to prevent Palestinian refugees from returning to their homes in areas that by then had officially become Israel. Prior to 1948, Jewish and international travelers in Mandate Palestine left numerous accounts of the villages there along with their inhabitants; the British cartographic surveys were only accomplished through extensive fieldwork in concert with local Palestinians to determine the names of towns and geographic features (Abu El-Haj 2001). After 1948, Amiran traveled through the land and viewed it primarily as a natural environment. He thoroughly documented temperatures, the heights of hills and mountains, and the location of rivers and springs. Even so, that landscape could only be seen as empty and natural once the vast majority of its Palestinian population had been excluded from it.

Two Incomparable Exiles

Both Hadawi and Amiran were forcibly pushed out of the places of their birth, and both traveled extensively throughout their lives. Professionally, both scholars saw themselves as empirical observers of the geographic scenery and had similar ways of framing their methods and goals. Yet they continued to be divided by their access to a state—a disparity that reveals the power imbalances at work throughout their lives. These imbalances are evident in the ways that Hadawi and Amiran were affected by landscape economically, through their ability to shape the lands in which they traveled, and socially, in terms of why they moved how they were treated as they moved. While Hadawi spent his time attempting to convince the United Nations to restore Palestinian property that had been lost, Amiran traveled to obtain funds to support modern development in Israel. At the same time, Amiran consistently benefited from the control that came with establishing new state institutions with the help of international aid. He thereby helped institute the disciplinary and economic infrastructure that would allow Israeli scientists to set up future funding bodies of their own.

Indeed, high levels of foreign aid and loans enabled the wholesale rationalization and modernization of a landscape that rapidly became Israeli after 1948, as examine in more detail in chapter 2 (Razin 1993). This was in combination with the sizable economic assets obtained through the seizure of Palestinian land, including fruit orchards, arable fields, urban built-up areas, beaches, and archaeological sites that double as tourist destinations. They were so valuable that Hadawi (1957, 31) could cite at least one source, Don Peretz, who argued that "abandoned property [i.e., property left behind by Palestinian refugees] was one of the greatest contributions toward making Israel a viable state." As I discuss in chapter 4, access to a state also impacts the possibilities for producing knowledge. For example, Rashid Khalidi looks at how the exodus of Palestinians between 1947 and 1949, and continued absence of an internationally recognized Palestinian state, has affected the archival records that are available for historic Palestine. Khalidi (2006, xxxv–xxvi) notes the "basic asymmetry with respect to archives is a reflection of the asymmetry between the two sides" of the conflict. Similarly, Salim Tamari and Elia Zureik (2001) explore the efforts to collect and digitize the archives for Palestinian refugees—archives that were dispersed with the refugees themselves.

In contrast, although Hadawi was also internationally funded as an employee of humanitarian agencies, his money came not as personal assets, loans, or aid among nations. Instead, it took the form of humanitarian aid from international bodies like the United Nations—bodies that ultimately controlled the money's direction and flow. In keeping with these more nebulous funds that were available to him, the institutions that he founded in exile were more transitory in terms of both location and duration, and this despite the best efforts of Hadawi and his associates. Partly as a result, Hadawi was often on the move as well, living in Lebanon and the United States, among other countries, before settling permanently in Canada (Bland 2004; Safieddine 2004).

Rather than being recognized for his effectiveness, Hadawi was forced to resign from his position at the United Nations in 1955 after he asked formally for economic restitution for the home that had been taken from his family after 1948. His resignation was forced because of a perceived incompatibility between his identity as an international observer, on the one hand, and on the other hand, as a refugee who had a right to compensation under international law (Fischbach 2003, 252–258). Hadawi would never see Jerusalem again, passing away in Toronto in 2004 at the age of one hundred. Hadawi was a British citizen, and thus he had the option of visiting, but not residing in, Palestine and Israel, yet he refused to do so out of protest (Fischbach 2003).[15] Amiran, in contrast, was widely recognized and celebrated for his work. His identities as a Jewish Israeli and international scholar were widely assumed to be compatible. Amiran increasingly was received as an equal among international power centers, whereas Hadawi's contributions to history only began to be appreciated decades later. The practical effects of statehood on the ground served as the fulcrum of the imbalance of power between the two scholars—an instance that will be seen time and again, albeit in differing guises, in the chapters that follow.

From Traveling Theory to Traveling Ethnography

If Hadawi and Amiran were influenced by the times and places in which they worked, then this book inevitably is also a product of my position as an international scholar. It is grounded in over six months of ethnographic fieldwork and archival research in Palestine and Israel plus one further month of archival research internationally. These periods include five months of participant observation at one of the premier Palestinian cartographic nongovernmental organizations (NGOs) in the West Bank as well as over fifty semistructured interviews with cartographers, urban planners, and civil engineers. In addition, I participated in more than thirty field visits with cartographers and field researchers throughout the West Bank, and took upward of ten thousand photographs of maps and documents in related archival collections in the West Bank, Israel, and the United Kingdom.[16] Throughout, I sought a representative sample of interviewees employed in the dominant cartographic institutions and organizations in the region. The combination of textual and ethnographic sources enabled me to explore the influences that shaped the entire process of making maps, from the initial setting of the grid and data collection—both gathering existing sources and making new observations—to the design of a particular map.

During my fieldwork, I was repeatedly reminded of how my position as an international scholar opened specific routes while foreclosing others. My US citizenship and other forms of privilege made it possible for me to work in both central Israel and the West Bank. This is significant given that many Palestinians experience routine discrimination when traveling between the two, even when their visas permit it, and Israeli citizens are formally barred by the Israeli government from traveling to the Palestinian Territories without special permission, although small groups of activists continue to do so. However, also as an international, my stay in the region was dependent on me leaving every three months, and I could be refused reentry each time I entered or returned. Such a refusal is routinely enforced without warning on Palestinians in general as well as North American and European scholars who are, or are believed to be, of Arab heritage.[17] Thus it is not infrequently used against professionals, educators, activists, and interns at international NGOs, despite the privileges conveyed by their US or EU passports (Abdel Fattah 2013; ARIJ 2013a; Miller 2013). During this research I was never refused entry, in no small part because of the fact that guards perceived me as a white international entering as a tourist, due to my Jewish heritage and last name through my paternal grandfather, to visit friends and family. My somewhat-extensive range of movement was granted therefore largely on factors of race, gender, sexuality, class, and religion as determined in part by appearance. It was still precarious, though, since it could be prohibited at any moment, at any checkpoint.[18]

This book is the result of an attempt to mobilize my privilege to highlight the role of international knowledge in the occupation, thereby pointing out how all knowledge is geographically situated. I sought to follow Edward Said's *traveling theory* to its logical conclusion, by conducting a traveling ethnography. Said (1983, 161) famously proposed traveling theory as the study of "what happens to a theory when it moves from one place to another." In his critique of the totalizing aspects of Michel Foucault's theory of power, Said introduces an analysis that might provide "attractive alternatives" to any theory that "left to its own specialists," would tend "to have walls erected around itself." As he outlines the goals of these alternatives, Said (ibid., 177–178, emphasis added) makes prolific use of geographic metaphors: "[They are intended] to *measure the distance* between theory then and now, there and here ... to *move* skeptically in the broader political *world* where such things as the humanities or the great classics ought to be seen as *small provinces* of the human venture, to *map the territory* covered by all the techniques of dissemination, communication and interpretation."

Said's borrowing from geographical conceptions of space in turn implies a recognition of theory's own material form—a form that can mutate when those theories move and change. He is not alone in his use of spatial metaphors in the study of theory. In the introduction to *Primate Visions*, for example, Donna Haraway (1990, 10, emphasis added) invokes Said's *Orientalism* to argue that primatology works through the "negotiating of *boundaries* achieved through ordering differences" that "mark off important social *territories*." Said and Haraway thus draw attention to the central yet often underappreciated role of space and geography in conceptions of knowledge. In so doing, they prefigure later notions of *translation* in Actor-Network Theory (ANT) as the way that actors can transform themselves in order to be incorporated into a network (Latour 1999, 15–16).

Yet to study such transformations, it is also necessary to pay heed to their negations and omissions, to the territories they make, and that make them, as well as to the paths of greatest resistance—the places where theory fears to tread. This involves extending Michel-Rolph Trouillot's (1994) accounts of the "silences of history" so that it includes the silences of geography.[19] In addition, even as Said critiques Eurocentric notions of colonized regions as empty, blank spaces, at times he maintains an artificial separation between cultural space and objective or physical space, with a focus on the movement of individual authors or schools of thought. In a similar vein, although the goal is an analysis of objective knowledge, to date ANT research has focused more explicitly on actors than on the spaces that they make, and through which they move. By contrast, this book is a combination of geographical, postcolonial, and STS perspectives that critiques theory and knowledge in concert with a critique of geographic space. It draws on existing work on knowledge-making practices while avoiding the widespread assumption that space itself is merely the background for the travel of theory and formation of networks.[20]

Theory makes and unmakes geographic passages through its movement. Passages can make and unmake theory.[21] In order to feel out these simultaneous affirmations and negations while taking their materiality into account, I conducted the better part of my research on the road. I also routinely strayed off the road, both literally and metaphorically, in an effort to materialize my attempts at scholarly reflexivity and better understand how the study of traveling theory alters the subjectivity of the traveling scholar in ways that might further nuance the resulting research. For similar reasons, in the conclusion to chapter 5, I present the notion of *refraction*, or reflexivity, with respect to spatial and material situatedness. In the spirit of refraction, and as is most evident in chapter 6, I do not restrict myself only to the kinds of dominant academic literatures that albeit misleadingly, are said to originate in Europe and North America.

This traveling ethnography began with the simple notion that in order to study traveling theory, it is necessary to travel with it. In an effort to mobilize my privilege to allow for an alternative rethinking, though one necessarily rooted in my own positionality, I made daily trips back and forth across the many boundaries that riddle the West Bank, using as many forms of transportation as possible. These included the segregated Palestinian and, alternately, Israeli buses and minibuses, taxis, UN jeeps, private cars, bikes, hiking trails, and at one point, a small red scooter. I sought out alternate routes, taking Israeli public buses into settlements, and then walking across to adjacent Palestinian towns, then coming back through checkpoints on foot. With the help of local activists and guides, I climbed through the ruins of the demolished Palestinian villages that are buried beneath the "green belt" of forests that line the highway that runs from Tel Aviv to Jerusalem—forests that were planted in part to surreptitiously hide those same ruins and secure Israeli government ownership of their land (Cohen 1993, 109). I traveled north to the Lebanon border and south to a Bedouin town in the Negev Desert, which at the time of this writing, has been demolished at least forty-nine times (Kestler-D'Amours 2013). While there, I helped to plant olive trees, but was reminded of the futility of our efforts when one of the saplings fell over and a teenage girl said, pointing to the Israeli soldiers and bulldozers on a nearby hill, "Don't worry. The trees will all be gone by tomorrow."

I also made regular trips along the segregated road for Palestinians, recently built with US funding. The road takes perhaps the longest-possible route between Bethlehem, south of Jerusalem, and Ramallah, to Jerusalem's north. Skirting Jerusalem entirely, the road runs on steep inclines and descends through the aptly named *Wadi al-Nar* (Valley of Fire), taking one to two hours to go between two cities that would be no more than twenty minutes apart by car on a direct road. I walked along the Wall, part of the broader Israeli Separation Barrier that consists, in places, of barbed wire fences surrounded by trenches and a broad exclusion area on either side, and in others, of an actual eight-meter-high concrete wall dotted with guard towers (see figure 6.1). Overall, it is built with brutal efficiency within the populated areas of the West Bank, on top of a crisscross of streets, homes, and fields—a route that is alternately eerily tranquil and suffused with suffocating teargas clouds.

As noted above, all this travel was thoroughly conditioned by my privilege as a white international academic. The hope was that by crisscrossing those boundaries in such innumerable ways, I might come to a more embodied understanding of how societies can be so intricately interwoven and also thoroughly disjointed.[22] Yet this privileged mobile life was itself circumscribed by the near moratorium on international travel to and from Gaza, and as such, it too was geographically curtailed in ways that reverberate throughout the research. The restrictions certainly do not prevent all research on Gaza (see Bhungalia 2009, 2012; Salamanca 2011; Tawil-Souri 2011b), but it does make intensive fieldwork there exceedingly difficult for most researchers—including locals whose lives are continually disrupted. It is possible for internationals to enter Gaza with special permission, often as representatives of humanitarian organizations. But whereas I walked long distances in the West Bank, and used a variety of public and private forms of transportation, in Gaza movement can be tightly controlled, making it very difficult to do precisely this type of traveling fieldwork. As such, it was not, for me as a researcher during the period of my fieldwork, a viable situation for studying competing empirical claims on the ground. Gaza remains a major crux of the conflict and requires much greater attention in the academic literature than it so far has received. Sadly, the dearth of research on Gaza is due in no small part to its very rigid closure by the Israeli military.

The choice to circulate across and through the landscape required giving up on becoming fully immersed in any one place. At times I became thoroughly disoriented, not sure which language I was supposed to be speaking, or to whom. I was almost vigorously thrown out of at least one taxicab for speaking the "wrong" language. Yet this living between societies did alter my views in small but innumerable ways. Such an approach certainly is not appropriate for all studies, and the potential forms of alternative travel are vast and have varying political and social implications.[23] Nonetheless, this account seeks to demonstrate that work that is physically conducted on the margins is a necessary part of wider projects to research the relationships between knowledge and power, including geographies of occupation that depend on the enforcement of strict, if ambiguous, boundaries.

Of course, the ways researchers move about are also central due to the potentially negative consequences of the work of international academics in boundary communities. For example, the continued influx of (temporary) international researchers into Palestinian refugee camps in Lebanon may have actually increased distrust. Moe Ali Nayel (2013) relates one incident where a female resident, Um Muhammad, asked a US professor to help her seriously injured son, only to have the professor reply, "'We will include your son's story in part of the study we are doing, and it will be published by Harvard.' Then, the professor asked me [Nyel] to tell anxious Um Muhammad that Harvard is an important university and when the report was published many people would read it." The apparent hubris and callousness of this response demonstrates that, while traveling complements existing studies where scholars embed themselves in particular communities, travel alone is not sufficient and suffers from its own set of pitfalls.[24]

Drawing the Outlines of this Book

In order to allow for the study of those pitfalls and related concerns, this chapter has illustrated the impact of colonial contexts along with their geographic divisions through an account of Hadawi and Amiran. In addition, it has sought to elucidate the method of traveling ethnography that is woven throughout this entire book. The ensuing chapters treat emblematic periods in the history and geography of the region. The case studies begin after 1967, amid the mid-twentieth century's broad attempts at mapping rigidly defined territories. They then move to the mid-1990s and after, to the increasingly compartmentalized contemporary efforts, including the intensification of microscale conflicts over sovereignty in the West Bank. Chapter 3 focuses on Israeli state population maps in the period between 1967 and 1995, which stretches from the beginning of the Israeli occupation of the West Bank and Gaza Strip, and into the post-Oslo period that led to limited Palestinian rule in those territories. Chapter 4 looks at the Palestinian state-building maps of the late 1990s and early 2000s, prior to and during the Second Intifada, or uprising, against Israeli rule. It examines nascent Palestinian sovereignty in the West Bank to understand how its concerted efforts are shaped by colonialisms past and present. Chapter 5 revolves around the changes that followed the formal end of the Second Intifada, which saw both increased mobility in the West Bank and increased segregation as the Israeli settlement project continued to expand. Chapter 6 synthesizes the preceding chapters on empirical cartography. Returning to the themes of this first chapter, it suggests that the reflexive movement of traveling ethnography must also involve drawing on more expansive literatures and sources. With this in mind, it carefully and critically incorporates the work of literary and political figures like Mahmoud Darwish, Audre Lorde, and Nawal El Saadawi.

In chapter 2, I argue for the benefits of studying the geographic production of knowledge while providing the theoretical context for the case studies' combined use of geography, postcolonial theory, and STS. After 1948, the lands of Palestine and Israel were developed in ways that privileged narrow forms of rationality above other concerns—and that paradoxically, led to a contemporary situation that sometimes appears to be thoroughly irrational. To better understand the broader ramifications of these changes, I begin by exploring postcolonial studies of cartography and empire in concert with critical geographic literature on the production of urban landscape. I then bring together STS with another subdiscipline of geography, critical GIS, in order to further contextualize the symmetrical framework of the book's overall analysis of cartographic practice.

As noted, chapter 3 focuses on Israeli state population cartography. The Israeli census is conducted by the CBS, which periodically carries out a direct count of the total number of inhabitants of Israel and collects statistical information related to population. As such, the census has played a central role in a conflict where demographic fears are paramount. Population maps are traditionally one of the only forms of professional maps that display people instead of geographic features like tracts of land. They involve the heavy use of statistical methods, and were one of the first types of maps to be made on a computer. With this in mind, I investigate the alternating inclusion and exclusion of Palestinians on Israeli census maps between 1967 and 1995. The analysis demonstrates how even the most abstract mathematical facts are conditioned by the landscapes where they are produced. Through an exploration of the geostatistical methods developed by Roberto Bachi, who for many years was the head of the CBS, I show how the types of math that Bachi used were altered by his twin goals of making scientifically accurate maps, on the one hand, while avoiding indicating Palestinian communities, on the other.

Even though Palestinians do not always appear on Bachi's maps, by many accounts there were significant numbers of Palestinians living in the areas that Bachi was mapping. This in turn affected the types of maps that Bachi made. But rather than a case of political concerns leading to "bad science," Bachi's methods themselves took a different tack, developing in a different way *within* the standards of science. In the process, those methods were intrinsically shaped by political concerns, albeit in ways that still allowed them to be considered fully scientific and statistically rigorous. One effect of this shaping was that Palestinian communities began to show up on maps as clearly delineated, albeit blank, areas. So Palestinians were made visible to Israeli statisticians through their detailed and violent erasure from the map.

In chapter 4, I examine more recent Palestinian state cartography. I investigate how past and present colonial legacies circumscribed the scale and extent of geographic data amid attempts to practice *sumud*, or steadfastness against the occupation, in the post-Oslo period (Allen 2008). While studies of the occupation since the year 2000 have often focused on the restriction of mobility, I argue that *stasis*, or the ability to stay in place, is also a central goal of Palestinian efforts. Curtailed mobilities indeed hamper Palestinian cartographers working in the West Bank. Equally important, however, are restraints on their ability to stay reliably within portions of the territories, without experiencing aggression from the military or settlers, in order to set up surveying equipment and take more extended readings. Stasis has had negative connotations for colonized peoples, who are mistakenly depicted as being stuck in place and time (Fabian 2002). Yet this negative view contrasts with the work of many in Palestinian social and political movements, where attempts to build sumud have sought to reclaim stasis as a form of enrichment, flourishing, and community building. However, these efforts have been resolutely opposed by the occupation forces. Israeli military aggression against Palestinian scientists has worked in tandem with raids against PA infrastructure and databases. These are further compounded by a shift in the locations of institutions and organizations away from Jerusalem, due in part to the building of the Wall and restrictive permit regimes. Such challenges to Palestinian stasis combine with legacies of British colonial cartography to intrinsically shape the content of PA maps in ways that are not visible simply by analyzing a completed map. Instead, they require delving into the very constitution of map data and practices to show how challenges to stasis also can lead to more transitory and precarious maps.

Chapter 5 includes an exploration of alternative urban cartography in NGOs. I compare the recent empirical maps of Israeli West Bank settlements made by one Palestinian and one Israeli NGO in order to demonstrate how intricately segregation influences the process of collecting, classifying, and visualizing geographic data. Israeli settlements have been built in the occupied West Bank with the express purpose of wreaking "infrastructural violence" (Ferguson 2012; Rodgers and O'Neill 2012; Salamanca 2014) by complicating attempts to separate the area from Israel so that it can form part of a distinct Palestinian state. They thereby adversely affect Palestinian urban development (Abdelhamid 2006) while pragmatically reinforcing the current situation where many Palestinians have neither formal Israeli citizenship nor a fully internationally recognized nation of their own. The two organizations in chapter 5 seek to observe the locations of particular buildings, roads, and fences in settlements that cartographers, for their own safety, may only be able to view from a distance. They employ similar cartographic methods with related goals, yet their maps don't always agree. For while Israeli efforts take place largely *within* Israeli settlements, thereby depicting Palestinian areas from *without*, for Palestinians it is the reverse. Palestinians work *within* the Palestinian West Bank, but must map the same settlement from *without*. In theory, this difference should not shape the data collected, but in practice, it has a profound effect on the resulting maps.

The differences in the two NGOs' maps show how segregated landscapes not only separate populations but also reinforce exclusive forms of identity, and reproduce disjunct observations, among cartographers who map areas, and use technologies that ostensibly are the same. The implications that result are explored more fully in chapter 6, where I study the broader ramifications of material hierarchies of knowledge production. Like empirical cartography, critical social theory has been shaped by, and narrowed within, specific (Eurocentric and Anglocentric) histories and landscapes. As a result, even postcolonial theory draws many of its insights from elite scholars trained in Europe and North America, as well as Australia and New Zealand. This serves to legitimate those theories, but sometimes it does so at the expense of the work of activists and others who help to make this kind of research politically and socially possible.

This book intervenes in debates over the role of international scientific knowledge by demonstrating that no single group has an ultimate monopoly on the facts that are accepted as true. This theme is next taken up in chapter 2, through its discussion of the role of colonialism in cartography. Far from erasing or overcoming the legacy of colonialism, digital tools have reconfigured and in some cases vastly extended the ways that maps can be used to enforce imperial control.

2 The Materiality of Theory

Maps have long been an intrinsic part of political and military struggles. Some of the earliest topographic maps of historic Palestine were produced by Pierre Jacotin, a senior officer in Napoléon's army, who followed the general's trail of battlefields during his failed invasion of the region (Godlewska 1999, 77–78). In the end, however, Jacotin's results were inconsistent because he could only work in sites where Napoléon's forces had won (Khatib 2003, 211). This tight coupling between cartography and control over land persists in a modified form in contemporary Palestine and Israel. It is evident in a recent text by Haim Srebro, the director of the Survey of Israel (SOI), where the author depicts cartographic surveyors as "pioneers." He claims that surveyors are the first in the field, "the first to thrust a stake in the ground. When they arrive ... the field is usually bare, often unapproachable" (Srebro, Adler, and Gavish 2009, 7). Srebro's words evoke images of unmapped land as uncontrolled space, as both empty and inhospitable. Such a depiction would have been all too familiar in Jacotin's day. Indeed, pioneers are often innovators, and innovation is closely associated with technological invention. But additionally, pioneers long have been the harbingers of military invasion and displacement. The persistence of pioneer metaphors thus points to the need to better understand how colonialism and empire continue to shape digital cartography.

The intertwined histories of maps and struggles for territorial control are also broadly relevant to the spatialities of knowledge. Recently, authors in a variety of disciplines have begun to refer to a "spatial turn" in social theory. Metaphors of mapping are widely used to refer to textured understandings of complex social phenomena (Law and Mol 2001; Warf and Arias 2009). Nonetheless, beyond specific subdisciplines of geography and, with quite-different implications, certain branches of mathematics where functions are referred to as *maps*, to date there has been less critical interrogation of mapping practices.

In order to contextualize this use of maps more fully, this chapter provides an analysis of cartography as both a political and scientific practice. Told through a narrative of how cartography has been embedded in the landscapes of Palestine and Israel over time, this chapter has two main goals. First, it serves to demonstrate how studies of the region, and research on the political impacts of technology more broadly, can benefit from a colonial and postcolonial frame (Abowd 2007; Abu El-Haj 2001). Constrained by national borders, nationalist histories of technoscience often attribute change only to causes that arise from within the nation-state. In contrast, a broader colonial and postcolonial frame allows for a study of how the space of the nation-state along with related notions like *Palestinian science* or *Israeli science* are themselves produced. It is, as Timothy Mitchell (2002) has noted, a frame that permits an exploration of persistent colonial legacies, rather than one that implies that colonialism is solely a process of the past.

Second, this chapter shows why critical theory cannot bracket the content of technoscience but instead must delve into both its material and theoretical details. As I argue in chapter 1, without such attention to concrete practices, it is too easy to fall into the oversimplistic view that science and technology are politically neutral, and hence insusceptible to political control. Opposed to this, there is the equally problematic notion that technoscience is thoroughly and solely determined by politics or social power. Both views share aspects of the determinism that has been thoroughly debunked by critical scholars (e.g., Peet 1985; Wyatt 2008b). In the former case, technology drives society, and in the latter, it is the other way around. In contrast, analyses of scientific and technological practices show the intricate social and material constellations that simultaneously contribute to specific aspects of both continuity and change.

The study of both colonial legacies and technological practice are integral to a geographical analysis of the ways that scientific knowledge has been shaped by changing political, social, and economic landscapes. As with science and technology, however, it would be mistaken to assume that such geographies are either natural and politically independent, on the one hand, or solely the result of intentional social control, on the other. To this end, this chapter elaborates on how spaces themselves are produced in ways that are also interwoven with the discipline of cartography. I examine how space is reshaped through land management practices, and how that reshaping in turn influences contemporary ideas about land, people, and technoscience.

This chapter is divided into three main sections. The first begins with an introduction to the study of the landscapes of Palestine and Israel before providing the context for the theme of *internationalism* by researching the role of colonialism and imperialism in scientific knowledge, and cartography in particular. I then turn to the literature on the production of *landscape*. The second section provides an outline of changes in the land in concert with the related development and modernizing discourses of the 1950s and 1960s—discourses that connect historic colonialism to the present day. There were numerous attempts to rationalize space between the founding of the state of Israel in 1948 and the 1967 war. They were carried out both on the part of Israeli state agencies and, in the West Bank and Gaza Strip, under the respective Jordanian and Egyptian administrators. International ideologies of modernization thus were closely connected with colonial discourses about the alleged inferiority of local populations. They were deployed in modified forms by Israeli planners whose development efforts were increasingly effected through statistical, and ultimately digital, maps. If the study of landscape is an area where geography approaches STS literature, then STS literature on the field, in relation to the laboratory, is another. In the third section, I move to critical GIS and STS analyses of how more recent digital mapmaking built on past colonial and military technologies during the transition to computer mapmaking that took place in the mid- and late twentieth century. The chapter closes with a discussion of what the STS method of *symmetry* can contribute to elucidating the links between colonialism, development, and technological practice. I argue for an expansion of the notion of *symmetry* in STS to include constructed geographic symmetries, such as that between Palestinian and Israeli maps that organizes this book. As discussed in chapter 1, however, the purpose of geographic symmetry is not to conduct an unreflexive comparison or commensuration of Palestinians and Israelis. Instead, it is precisely to highlight the extreme imbalances of power and knowledge between them—due in large part to the ongoing Israeli occupation as well as the specific forms of continuing international involvement and aid.

Irrational Rationality in the Landscapes of Palestine and Israel

Landscapes serve to both alter and perpetuate colonial knowledge legacies. One such legacy is the narrow conception of rationality that has been legitimized through the use of digital technology—at times with bewildering results. In Palestine and Israel, the drive for rationality has led to an apparent irrationality in the land, an intricate twisting and turning of competing sovereignties, and all the practical issues that goes with it. These issues are the subject of daily conversations among travelers and commuters of a sort that is common elsewhere: which roads are open and why, where there are delays and for how long, and so on. Yet in the context of the conflict, which is euphemistically referred to in Hebrew as *ha-matzav* (the situation), such talk is incredibly elaborate and extensive—a replacement for the weather as the go-to topic of conversation. Traffic jams are both deadly serious, as when Palestinian ambulances are prevented from entering East Jerusalem to go to nearby hospitals, and the subject of continual satire. The satire arises in part as a way of coping with brutal segregation that is neither "simply" irrational nor an inscrutable force of social nature. Instead, rationality and irrationality, common sense and nonsense, frequently go hand in hand.

Despite how it is sometimes represented, the ongoing conflict between Israel and the OPT is not the result of some ancient and illogical sectarian quarrel. It instead emerged in concert with international methods of developing and analyzing land—methods that privilege specific forms of rationality and quantification. Many of the results of their methods can be seen in the administration of the Palestinian Territories following the start of the Israeli occupation of the West Bank and Gaza Strip in 1967. Prior to the 1967 war, land management efforts were aimed at clearly defining distinct and separate spaces for Palestinians and Israelis. This changed after 1967, when the Israeli government began attempts to enforce a single state in both Israel and the Palestinian Territories. The OPT thus were *irrationalized*, or made irrational, as increasingly intricate and obscurantist governance procedures began to overlap with, and not infrequently contradict, one another. At the same time, however, the government refused to formally annex the Palestinian Territories to Israel, in a move that would have put them under the jurisdiction of Israeli civil law rather than the separate (and thoroughly unequal) military codes that are currently in force (Barak 2005).

Such irrationalization resulted in part from a conscious effort on the part of the leadership of Israel. The goal was to solidify the Israeli position both through forced movement and brute territorial expansion as well as by more subtle means: namely, by creating conditions that are so dismal and inscrutable that they encourage Palestinians to emigrate. Rationalism is often allied with what is predictable and expected, and the irrational with the unexpected. However, those efforts also had unexpected results— not least, the strengthening of Palestinian resistance. Eyal Weizman (2007, 4) describes the proliferation of borders that resulted: "Against the geography of stable, static places, and the balance across linear and fixed sovereign borders, frontiers are deep, shifting, fragmented and elastic territories. Temporary lines of engagement, marked by makeshift boundaries, are not limited to the edges of political space but exist throughout its depth. Distinctions between the 'inside' and 'outside' cannot be clearly marked." By drawing attention to the ambiguity of land management practices in the (so-called) frontier, Weizman highlights the simultaneous inclusion and exclusion that continues to be evident in the Israeli development of the West Bank (Ophir, Givoni, and Hanafi 2009). This has given rise to a condition that Meir Margalit (2010, 23) has called, "permanent temporariness"—one that has been analyzed through Giorgio Agamben's notion of the *state of exception* (Shenhav and Berda 2009). Wendy Brown (2010, 30) has noted that such apparent irrationality may result from multiple shifting rationalities that, as we will see, have been made inherent in the landscape.

So although the West Bank is highly segregated, in practice it is incredibly difficult to demarcate precisely where the boundaries between such multiplicitous areas fall on the ground. The effects might be comical if they were not so debilitating. Palestinians are regularly tried for crossing borders that are functionally invisible and whose precise location, for "security" reasons, they aren't officially permitted to know. To give just one example, despite paying Israeli taxes for decades, Diana Kurd, a Palestinian widow from Anata, was stripped of her pension because her deceased husband had slept on the wrong side of their bed. Namely, his side of the bed lay *outside* the Jerusalem city boundary, while hers lay within it. Yet the boundary itself was determined according to the Israeli authorities' maps—maps that at such fine scales, are usually classified (Hasson 2011).

The borders that intrude into the bedroom also split towns in half and sever them from their surrounding regions. The construction of the infamous Wall within the West Bank has created enclaves in Barta'a and western Bethlehem.[1] Despite officially being part of the West Bank, these Palestinian towns are left on the Israeli side of the Wall, which functions as a de facto new border on the ground. The border has literally moved around them (United Nations Children's Emergency Fund 2008).[2] It is as if Canada built a wall around northern Vermont and then told the residents that they were all now illegal squatters on Canadian soil. Physical inclusion into territory annexed by Israel, however, does not come with citizenship or egalitarian political inclusion in the Israeli state. Instead, the Palestinian inhabitants are trapped. Because of the Wall, they are physically unable to enter the West Bank, but legally they are equally unable to enter Israel even though, for all practical purposes, that is where they now reside. Their presence has become logistically and legally impossible—but partly out of sheer courage and determination, there they are still.

In this context, it is no surprise that some of the clearest maps of East Jerusalem alone show over a dozen relevant borders (figure 2.1). Those borders are thoroughly intertwined with attempts to count and manage populations. Helga Tawil-Souri (2012) argues that over time, the occupation authorities have focused less on efforts to control specific pockets of space—perhaps in part due to the proliferation of borders. Instead, there is a move toward tracking specific individuals through a multifaceted regime that allows varying degrees of access. Through this process, blatant discrimination is obscured by a proliferation of categories and differently colored ID cards that hide the broader coherent rationality at work.[3] This segregation by individual rather than territory also parallels a shift in GIS, as discussed more fully in chapter 3. Over time, GIS mapmaking practices have shifted from the mapping of wholesale territories to the analysis of large data sets for populations of individuals (Longley and Batty 2003). This too can serve as a form of apparently irrational obfuscation, even as it is a consequence of the maliciously clear and undifferentiated, one-size-fits-all treatment of large populations.

Such apparent irrationality can result from any number of issues, from the obdurate legacies of historical forms of reason, to purposeful obfuscation, to the mobilization of irrationality and ambiguity as strategies

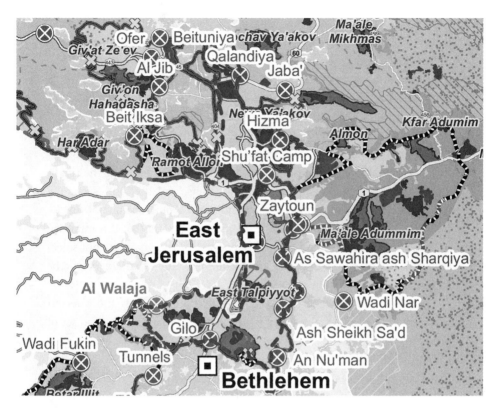

Figure 2.1
The intricacy of the geopolitical situation can be seen in this map of East Jerusalem, which lists over one dozen borders and barrier types for that region alone. This detail is excerpted from UNOCHA 2012a, S2. The map is the responsibility of the author. For updated UN maps of Palestine and Israel, see http://www.ochaopt.org/mapstopic .aspx?id=20&page=1 (accessed April 2, 2016).

of control. To better understand such mutating logics, it is necessary to explore how logic and rationality became the goal of cartography and related forms of land management in the first place. To that end, I now turn to the role of rationality in the broader intertwined histories of cartography and empire.

Making Empires, Making Maps

Maps have long been considered to be scientific representations of territory, but recently geographers have begun to question whether either of these characterizations—as purely *scientific* or a *representation*—is entirely deserved

(Kitchin and Dodge 2007; e.g., Latour and Hermant 1998). Both notions historically have involved a clear separation between maps and the rest of the world. Jeremy Crampton (2001, 693) has outlined how, in the 1950s, the view of maps in professional cartography also was predicated on "a clear separation between the cartographer and the user." It was believed, Crampton notes, that maps that function as *communicative* intermediaries between the representation, the map, and the physical world could bridge this separation. But that communication model implies that maps are not themselves material, whereas more recent critical scholars argue that maps and the material world—including maps, users, and the land—are fully imbricated. Thus a map is not a representation, a pure abstraction in the traditional sense, because it is not exiled from the world that it is intended to represent. In addition, scholars have contended that maps are not solely scientific in the sense that the knowledge that they contain might be considered purely objective.

Crampton and John Krygier (2006, 12) summarize this more recent scholarly view of maps as the recognition that a "map is a specific set of power-knowledge claims" that itself has a material form (see also Dodge, Kitchin, and Perkins 2009). Yet those claims do not need to represent the Earth as it currently is. Instead, they might suggest what the world could be, or what the mapmakers want it to be. This is the case of Israeli state maps printed in the 1960s. On those maps, Palestinian villages were depicted as ruins in anticipation that they one day would be so, since the Israeli government continued to prohibit Palestinians from returning (e.g., SOI 1964). Thus, maps not infrequently portray a world that the mapmakers plan to make rather than one that is already evident. As a result, John Pickles (2004, 93) argues that "mapping, even as it claimed to represent the world, produced it."

However, the representations of what *the world* should be were hardly the result of idle imaginings. Instead, they were made with the intention of pleasing heads of state—monarchs who had the power to realize those dreams. The belief that the sovereign could both make and unmake territory through maps is illustrated in a tale told by Eduardo Galeano. Galeano (1997, 149) notes that during a diplomatic row with Bolivia, "an infuriated Queen Victoria supposedly called for a map of South America, chalked an X over Bolivia, and pronounced sentence: 'Bolivia does not exist.'" To Queen Victoria, drawing an X on the map was equivalent to erasing the country on the ground from existence. Indeed, as a monarch, this may not have been simple hubris. The fact that many maps were often produced for sovereign rulers complicates the maps' pretention to depict an already-existing reality. It wasn't prudent to posit entities that didn't fit the ruler's conception of what should be. So rather than focus on current conditions, cartographers regularly depicted territories that were favorable to the ruler who commissioned the map—territories that were in the process of becoming actualized, such as an expanded realm. This relationship between maps and sovereign power suggests that maps are a useful means to describe the Earth *precisely* because the Earth was made and remade in the maps' image, not the other way around (Harley 2001; Pickles 1995, 2004).

This view that "maps make space as much as they record space" (Crampton 2010, 48) first emerged in the literature on critical cartography. Critical cartography gained salience in the late 1970s, but it dates its antecedents to the years after World War II and the work of John Kirtland Wright (Crampton and Krygier 2006; Edney 2005; Pickles 2004). A key subarea of research involves demonstrating the confluence of political and scientific designs in the role that cartographers played in European empire from the fifteenth to nineteenth centuries. Yet over time, these imperial ideologies would transform into discourses of modernity. Related discourses were employed around the world in the development efforts of the 1950s and 1960s—namely, during the wholesale transformation of the landscape of Israel after 1948. The geographical subdiscipline of critical cartography includes both the historical studies of cartography and empire that are examined in the next section as well as the critical GIS work discussed later in the chapter that analyzes contemporary digital mapmaking.[4] This chapter therefore explores the colonial roots of modernity, including how a series of vast and varied empires have been connected across time and space through the figure of the map.

Cartographies of Empire and Empirical Cartography

Benedict Anderson (2006) famously argues that capitalism and print culture originally led to the rise of national identities. Anderson describes one counterpart of this process in Southeast Asia, where the colonial state and dissemination of maps helped to make it possible to create national disciplines like archaeology. Thus he points to the tight links between colonialism, maps, and the institutional conditions of possibility for science. This has been taken further in geographical work on colonialism and one shift in conceptions of cartography. From a skill with artistic undertones, like drafting or printmaking, over the course of the eighteenth and nineteenth centuries mapmaking increasingly became a profession whose practitioners sought to make objective and scientific claims. Matthew Edney (1997, 2) highlights the connection between cartography and imperialism when he claims that "the empire exists because it can be mapped" (see also Pickles 2004). Similarly, Mitchell (2002, 9) contends that "in recent years the production of maps has often been taken to epitomize the character of colonial power, and by extension the power of the modern state."

Over the years, critical cartographers have conducted extensive historical studies of how cartographers came to impose mathematical order on the earth, thereby helping to legitimize the appropriation of land (Livingstone 1992). In an account that resonates with the tale of Jacotin's procession through historic Palestine that opened this chapter, Thongchai Winichakul describes how troops from Siam advanced in tandem with British surveyors in an 1885 conquest. As part of a military campaign, the surveyors and the kingdom's forces leapfrogged one another across the land. This process enabled the initial consolidation of the country that is now Thailand. Winichakul (1994, 82) notes that "sometimes, mapping advanced one step ahead of the troops. Then the military followed, making the mapping proposal of the areas come true." He maintains that only the military "could provide the authority under which mapping could be executed. In a sense, map anticipated the space; force executed it; map again vindicated it."

Cartography was able to vindicate conquest in part because it was increasingly viewed as an empirical science with practical import to the state. In the case of the British colonization of India, Pramod Nayar (2012) has remarked how "discovery" amounted to a form of "narrative possession" that later fed a shift from discovery, as a sense of wonder, and toward a quest for specific information. This was precisely the kind of information that would come to be known as *factual* or *empirical* descriptions, supposedly value-free statements of objective knowledge acquired and standardized from the senses, and vision in particular. It is the type of knowledge that is invoked and legitimated in simple calls to "Come see," or "Just look for yourself," as noted in chapter 1. As the cartographers' trade morphed into a profession, and an empirical one, mapmakers began relying on field surveys to observe local conditions. They also started to believe that their knowledge was universal, stretching beyond those local conditions (Latour 1987; Turnbull 2000).

In his analysis of the rise of cadastral maps in Britain, James C. Scott stresses the geographic importance of this striving for universality in contexts where outsiders sought territorial control. He argues that the map's value "lies in its abstraction and universality. In principle, at least, the same objective standard can be applied throughout the nation, regardless of local context." Overall, explains Scott (1998, 4–45), land maps "are designed to make the local situation legible to an outsider. For purely local purposes, a cadastral map was redundant. Everyone knows who held, say, the meadow by the river."[5] It didn't take long for the standards applied "throughout the nation" to also be applied in overseas colonies, albeit with varying results that required ongoing innovation. Nevertheless, despite such attempts at standardization and universality, Edney (1997, 17) points out that "the European knowledge of each empire is accordingly far more incomplete and nuanced than has often been recognized."

In the case of the British Empire in India, due to the practical concerns of operating in the landscape "the British could never implement the technological ideal" (ibid.) of triangulated surveys, leaving them instead to combine different surveys within a single grid in an attempt to construct a general map. Yet this apparently did little to dampen the drive to depict territories as if they were distinct, unchanging, and internally consistent entities (Edney 1997; Winichakul 1994). Indeed, Edney contends that the attempt to universalize cartography was so influential, in spite of its shortcomings in practice, that there were no substantial differences in appearance between maps of nation-states and those of empire. Nations and empires were seen as being part and parcel of overall territories of rule—a view that intentionally papered over differences between nation and colonies (Edney 2009). This was also the case under British rule in Mandate Palestine prior to 1948. Palestine was viewed as a core part of the British Empire more broadly (El-Eini 2006). The legacy of the attempts to standardize the British ordinance survey in the United Kingdom (Hewitt 2010) are clear in the British emphasis on a consistent triangulated survey over "all" of historic Palestine (Gavish 2005). Similarly, colonial officers were moved across different territories, including from Egypt to Palestine and beyond.

However, just as European cartographic methods varied over space, they also were never solely *European* methods. The history of dominant Western cartography includes recurring contributions from people around the world (Macleod and Kumar 1995), and Kapil Raj (2007) has pointed out the pitfalls of the "diffusion model" of science, whereby cartography was believed to have spread out from a central metropole to the colonies (Craib 2009). To this end, David Turnbull (1996, 2000), among others, has argued that all scientific knowledge is local knowledge, and not least when it comes to maps. Yet many of the broader contributions to geographic knowledge have gone uncredited and undervalued. Colonial and national maps frequently served to erase the presence of indigenous groups in part because members of those groups did not always seek to define rigid territories of the sort that made sense to the European mapmakers. Indigenous communities were thus left blank on maps, relegated to the realm of ethnographies and travel accounts (Safier 2009).

But this was not always the case. At times, imperial cartographers went to the other extreme—seeking to map local groups in keeping with intricate population counts and categorization methods that largely were imposed by the cartographers. Not surprisingly, there were ulterior motives. In Syria and Lebanon, French cartographers presented groups according to religious subdivisions as a means of dividing and fragmenting the Arab majority (Gratien 2013). Similar methods were not unknown in historic Palestine under the British. However, such practices extended far beyond the British and the French. In Siberia in the seventeenth and early eighteenth centuries, the numerous indigenous communities were depicted as a way of highlighting Russian unity and superiority by contrast (Kivelson 2009). Even so, in some cases official cartography explicitly incorporated indigenous practices, as in the Qing Empire in China where indigenous cartographic methods served to legitimate the ruling dynasty (Hostetler 2009). Yet whether indigenous people were erased or explicitly incorporated, maps were largely used for their subjugation, and as a means of extending imperial sovereignty into new areas.

Producing Landscapes for Traveling Cartographers

So critical cartographers generally agree that maps have their roots in empire, while also acknowledging that even dominant cartographic knowledge is far more heterogeneous than was first understood. Cartographies thus have their own geographies, including the ways that mapmaking practices vary over time and space. These are one subset of what I describe in chapter 1 as the *geographies of knowledge*, a concept that draws on a concern, in the literature on cartography and empire, for the intrinsic relationships between knowledge and landscape. Despite this, however, there are ongoing disagreements between scholars over the best way to conceptualize these geographies of cartographic practice. In particular, recent studies of maps and empire evince a tension between scholars influenced by French poststructuralist theory and those by Anglophone Marxist traditions. The poststructuralists study maps as discourses and texts (Harley 2001; Wood 2010)—or more appropriately, images (Cosgrove 2008)—to be interpreted. Researchers who draw on Marxist and neo-Marxist work in geography tend to analyze the practices, institutions, and power regimes that frame a map even before a single line is drawn (Cosgrove 2007).

This book represents an attempt to reframe and work through this divide in the literature. In some ways, the split parallels the much-maligned but ever-recurring arguments over structure versus agency in social theory. It also relates to debates in the history of capitalism and expertise more broadly. For example, Robert Jessop and Timothy Mitchell disagree as to the value of, alternately, producing broader evolutionary narratives that "seek to explain why only some economic imaginaries" are institutionalized at the expense of others (Jessop and Oosterlynck 2008, 115–116) and the perhaps more localized project of "tracing the specific history of a movement … the sites of economic knowledge it brings into being, the kinds of representation it makes possible" (Mitchell 2008, 1120). On the one hand, a search for evolutionary patterns may obscure other kinds of change and other types of knowledge. On the other hand, conceiving of knowledge solely in terms of sites or localities can efface broader injustices, including ongoing efforts to maintain systemic hierarchies of power.

Despite their theoretical differences, at times both Jessop's and Mitchell's accounts align systems of power with broad-scale narratives. Similarly, scholars routinely place structures in a simplistic opposition to a local knowledge that is conceived of solely in terms of contingent forms of agency. As the study of how to conceptualize spatial variation, critical cartography is uniquely suited to contribute to debates over the scale and place of critical research—debates that are central to social theory. Work on the geographies of knowledge thus provides opportunities for scholars to move beyond dichotomies between structuralist and poststructuralist accounts, while also pointing to similarities in structuralist and poststructuralist approaches.

As the name implies, study of the geographies of knowledge involves researching how forms, contexts, and contents of knowledge are constituted across space and time. It also can incorporate variations in broader epistemological concerns. These can include analysis of the very idea that knowledge exists as a separate category of experience and practice. It might also involve analyzing the differing values attributed to what, if anything, counts as knowledge in different societies. Yet the main idea remains that knowledge is not universal or everywhere the same, and that an understanding of how knowledge changes in different local, national, and international contexts is a worthy object of attention. The geographies of knowledge along with the related literature on *landscape* therefore provide one means for rethinking the implicit theoretical binary between oversimplified local particularities and equally oversimplified global patterns. The study of landscape allows for an explicit account of the nuanced and many-scaled spatialities (Martin 2011) that are at play.

So, far from being a system that was simply imposed wholesale on a pre-existing territory, cartography was developed *through* territory as part of a self-reinforcing process. Maps served to define specific territories, depicting them in precise ways, and the redefinition of those territories then served to reshape the practice of cartography. Imperial maps made by early scouts and explorers laid out routes of travel and trade, thereby facilitating colonization (Scott 2006). This led to maps of newly established forts and tracts of private property (Akerman 2006). The towns that were subsequently established then served as bases for further travel to collect data for yet more maps.[6] Colonial land management in this way afforded the conditions of possibility for cartography, and through the spread of urban planning and design, cartographic ideals would also come to further reshape the landscape. This was seen in the imposition of a grid pattern of streets in New York City in the early nineteenth century (Koeppel 2015) and in Georges-Eugène Haussmann's reconfiguration of Paris in the 1860s (Harvey 2005). In *Colonising Egypt*, Mitchell describes one such transformation in Egypt during the same period: the addition of Boulevard Muhammad Ali in Cairo, which was "ploughed diagonally through" hundreds of houses as well as "mosques, mills, bakeries and bath-houses." Mitchell (1991, 65) cites a

British commenter who likened the resulting destruction to "a city that has recently been shelled."[7]

Land management practices would be reconfigured over the course of the twentieth century, which saw both the persistence of empire together with the rise of postcolonial states, and the Middle East is one central area of debate over current imperial trends.[8] This was combined with the transformation, but not excision, of imperial legacies under resource management plans in postcolonial states. The plans were effected through modern geographical knowledge that built on colonial-era innovations with the aim of creating *visible* national geographies (Bhabha 2004; Mitchell 1991, 67). Shiv Visvanathan (1988, 264, emphasis added) draws attention to the situated visuality of the connection between modernism and colonial science, claiming that "modernity was a *vision* of conquest." Vision was emphasized and observation was believed to be superior to the other senses. As such, vision was seen as one of the primary means of creating knowledge that was both universal and objective. This nevertheless led to a neglect of the ways that the act of observation itself is situated in time and space.

The Production of Landscapes, the Production of Maps

Building on Henri Lefebvre's (2002) work on the production of space, the literature on the production of landscape provides an avenue for investigating the ways that geographic knowledges and related observations are situated in material landscapes (Massey 1994, 2005).[9] In recent decades, scholars in critical urban planning, development studies, geography, and anthropology have stressed the shaping and reshaping of the land (e.g., Domosh 1996). Neil Smith (2008, 79) has noted the resulting unity of concepts and material, albeit from a standpoint that privileges the material: "While the emphasis ... is on the direct physical production of space, the production of space also implies the production of the meaning, concepts, and consciousness." In some disciplines, this has been referred to as the *coproduction* of knowledge and materiality, which implies an understanding that there is a back and forth influence between society and knowledge, on the one hand, and the material world, on the other. Yet in recent years this has been taken further in disciplines like geography and anthropology in order to acknowledge there is no clear boundary between society and materiality, and that indeed they cannot be fully separated. In this vein,

Timothy Mitchell (2002, 6) argues that for social science, "the distinction between the material world and its representation is not something we can take as a starting point."[10] Instead, a fuller understanding is needed of how the idea and practice of this separation has arisen.

Research on landscape encompasses a number of literatures. It includes the work of Carl Sauer (1925) and his students on historical changes in North America as well as Cindi Katz's (2001, 1229) more recent call for studying *countertopography* as a way to "imagine a politics that maintains the distinctness of a place while recognizing that it is connected analytically to other places." Sharon Zukin (1991, 22) has called landscape "the cultural product of our time." In Palestine and Israel, Weizman (2002) has investigated the "folded terrain" of the West Bank in order to argue that "the conflict manifests itself most clearly in the adaptation, construction and obliteration of landscape and built environment."[11] Stressing production, Zukin and related scholars in geography (Domosh 1996; Harvey 1997, 2005, 2006; Katz 1994, 2006; Smith 2008; Soja 1996, 2000) coincide with Barbara Bender's call, based in archaeology, architecture, and the humanities (Misselwitz and Rieniets 2006; W.J.T. Mitchell 2002; Pullan and Baillie 2013; Sorkin 2002), to study the ways that landscapes are "tensioned, always in movement, always in the making" (Bender and Winer 2001, 3).

Of particular relevance to spatial concerns, Mitch Rose and John Wylie (2006) have argued against the "flattening" trend in recent social science, which has seen a shift toward *topology*, notably including the work of Bruno Latour (1993) and Agamben (1998, 2005). Like studies of landscape, social researchers working with topology tend to conceive of space as always mutable and changing. However, topology has its roots in geometry and set theory, and these are loaded with their own specific forms of calculative rationality not unrelated to the sort investigated here. Thus, while not averse to critical theories that draw on topological thinking, this book follows Rose and Wylie's (2006, 477) much broader call for the qualitative and quantitative study of landscape as a form of "dynamic materiality," a "tension of presence," and absence that "reintroduces perspective and contour; texture and feeling; perception and imagination."[12]

Even though *landscape* is a central term in geography, often it is defined only implicitly, through its use in a variety of analyses both within geography and beyond (Cosgrove 2008). This has resulted in a series of heterogeneous uses, most of which tend to privilege observation from above. Yet as Mark Dorrian and Gillian Rose (2003) contend, landscape may also be productively mobilized as an oppositional term. This is true particularly to the extent that the concept of landscape has the potential to encompass, and therefore allow for an exploration of, tensions between the material and subjective, the inside and outside, the past and present, and the representable and nonrepresentable. Rather than predicated on an a priori belief in any specific dichotomy (ibid., 16–17), such an account is itself concerned with the ways that colonial practices are ever anxious (Stoler 2010), obdurate, and pragmatic in a colloquial sense. Indeed, recent scholarship has shown how these dichotomies are partly effects of colonial geographic legacies and practices (Keane 2007; T. Mitchell 2002; Smith 2008). For example, in her account of imperial travelers, Mary Louise Pratt (2007, 61) refers to the "textual apartheid that separates landscape from people, accounts of inhabitants from accounts of their habitats." So Pratt views landscape as an effect of colonial attempts to divorce or deterritorialize indigenous human subjects from the places where they lived—attempts that didn't preclude romanticizing them as somehow primordially connected to the land.

Thus my efforts to analyze cartographic practice through conceptions of landscape are expressly not aimed at reinstating strict divisions between people and land, subjects and objects. Instead the goal is to respatialize technoscience, calling attention to how knowledges have been divorced from their contexts and allowing for greater attention to the varying spatialities of geographic production.

This book thereby contributes to the spatial turn in social theory and the humanities as well as a broad movement within STS, geography, and postcolonial theory that "locates these problems of colonialism, global expansion, and translation within the history and practice of science, rather than outside it" (T. Mitchell 2002, 7). It also questions the varying implementation of spatialized notions like the geographic *inside* and *outside* of scientific practice.[13] The study of the production of space allows for an exploration of how the very constitution of the scientific data is shaped by landscapes in ways that are far more fundamental and diverse than might be evident without concerted attention to geographic concepts. In the next section, I build on literature on cartography and empire, and the work on the production of space, in order to examine attempts to rationalize the landscapes of Israel and Palestine after 1948. Such attempts connected forms of colonialism with the contemporary digital technologies that are discussed in the following section. Their legacies, in turn, would ultimately enable the sometimes-absurd regimes noted earlier in this chapter.

The Rational Landscape of Palestine and Israel

Over the course of the twentieth century, cartography's imperial legacies would uniquely influence efforts to reshape the geographies of Palestine and Israel. In Israel in particular, efforts to fit the land to meet the map have dominated since before the state's inception in 1948. As described in chapter 1, the Governmental Names Committee's efforts to Hebraize place-names led to extreme toponymic transformations. Yet the land itself was also rationalized. Throughout, the wholesale reworking of the land aimed to make it legible, modern, measurable, and economically quantifiable.

For example, no sooner had the 1949 armistice been signed than work began on the initial stages of the National Water Carrier, a series of interconnected pipes, pumps, tunnels, and canals that would help further industrial agriculture in the region. They were designed in part by engineers from the United States, and the aim was to divert water from the Sea of Galilee toward the coast of Israel, bypassing the West Bank and preventing Palestinian access to drinking water (Alatout 2009; Davis, Maks, and Richardson 1980, 7–9; Fischhendler 2008). The importance of water infrastructure is evident in the fact that one of the first actions of the Palestine Liberation Organization (PLO) was an attempt to sabotage the National Water Carrier in 1964 (Wolf and Ross 1992, 931–937). But this is only one example from an entire host of similar projects, many of them funded from abroad. Across Israel in the middle and late decades of the twentieth century, masses of highways, forests, and large irrigation installations were wrought from infilled marshlands, dynamited hills, and the rubble of demolished Palestinian towns. Through these and related efforts, the attempt to "make the desert bloom" also built suburbia in Israel and Palestine.[14]

These large-scale infrastructure projects helped to stabilize the young Israeli economy while providing international legitimacy in the name of modernization. Many of them also led to environmental degradation.

Some initiatives, such as the draining of the Hula Valley, have since been partially reversed (Inbar 2002; Levin, Elron, and Gasith 2009; Watzman 1993). Yet concerns about the impact of modern development—for example, Elisha Efrat's (1994) caution with respect to the construction of highway 6's effects on Palestinian agricultural lands—have not halted construction. This suggests that a sense of inevitability continues to prevail in development circles despite ongoing opposition (Garb 2004).

The Israelis were not alone in their efforts. Between the 1949 cease-fire and the beginning of the 1967 occupation, the West Bank and Gaza Strip were under Jordanian and Egyptian control, respectively. Both the Jordanian and Egyptian governments undertook related modernization efforts in the territories under their jurisdiction. The Jordanians, for instance, attempted to demonstrate their legitimacy in Jerusalem by conserving and opening Christian religious sites to high-profile visitors such as the pope (Katz 2003). This changed after 1967, however, when large-scale investment was increasingly stunted in the Palestinian Territories due to the restrictions of the Israeli authorities, who now officially controlled them, and who instituted an impenetrable regime of building permits that are required though almost never granted for even the smallest modifications (Coon 1990; International Peace and Cooperation Center 2009; Tamari 2001; UNOCHA 2011a).[15]

National, international, and colonial (particularly British Mandate) modernizing discourses were all thoroughly intertwined during these transformations (Gasteyer and Butler Flora 2000)—as were financial motives (Levinson 2007). In Israel, the specific lineaments were combined in a fashion that was unique to Zionist discourse (Goldshleger, Amit-Cohen, and Shoshany 2006; Shohat 1992; cf. Gorney 2007), if also thoroughly tied to discourses abroad. In Jordan as well as nearby states like Syria, many likewise drew on international discourses, including complex admixtures of modernist, nationalist, and colonial tropes. This led to what Lisa Taraki (2008b, 63) has called a "uniquely Arab expression of modernity."[16]

Although unique, the Palestinian Territories also participated in a much broader postcolonial emphasis on self-sufficiency and a partial rejection of European hegemony. In contrast, Israeli modernizing discourses fed into colonial tropes in ways that were designed to appeal to power brokers in Europe and North America.[17] Zionist land management efforts traced their immediate roots to before the establishment of Israel, especially to the work of Zionist agencies in the 1920s to drain swamps like those in the Hula Valley. The draining, it was claimed, would heal a landscape that was supposedly "sick" from both malaria and economic neglect. This was accomplished without any reference to the fact that many areas were already actively managed by Palestinians in ways that may well have been more ecologically sustainable than the methods that replaced them (Gorney 2007, 267; Hirsch 2009; Sufian 2007, 138–144). In the process, Israeli geographers built on prestate discourses that alternately either depicted Palestinians as an integral part of the landscape or argued that they were "completely absent" (Nassar 2003, 150).

After 1948, Israeli geographers increasingly positioned themselves as cultural Westerners, casting local populations as unfit for development. In so doing, they bought into the broader perception of development as a linear process with defined goals that were exemplified in the societies of Europe and North America.[18] In the words of Moshe Brawer, Israelis accomplished "rapid and extensive urban and rural development, and provision of the modern efficient services and infrastructure of an advanced state." These achievements often were pitted against those of neighboring states, such as by claiming, as Brawer (2008, 153–154) puts it, that "the West bank had suffered from neglect, stagnation, and partial depopulation during its 19 years under the Jordanian Kingdom (1948–67)" in an apparent attempt to paint the 1967 Israeli occupation as if it were of benefit to the local population, and this despite ongoing Palestinian resistance.

Developing Colonialism and the Imperial Modern

Such colonial arguments were not solely relegated to the early and middle decades of the 1900s. Instead of disappearing, overt colonialism was subsumed into discourses of development. Under the guise of spreading modernity, this enabled the furthering of imperial forms of control both within and beyond the nation-state (Escobar 2011). In recent decades, prominent Israeli developers and academics have consistently presented themselves as colonial modernizers (interview 3, Israeli geography professor; interview 4, Israeli geography professor). One example is Srebro's characterization, at the beginning of this chapter, of unsurveyed land as both *bare* and *unapproachable*. Srebro implies that Palestinians have failed to develop the space—omitting mention of how they might be expected to do so under the economic and legal restrictions of the occupation—while also cluttering it up (with their presence) in ways that prevented it from being accessible to the "real" developers—namely, Israeli authorities and multinational corporations (e.g., see Srebro, Adler, and Gavish 2009).

This discourse leads to the continued effacement of Palestinians and alternative, including sustainable, methods of treating the land. In this context, Elia Zureik has called the attitude of Baruch Kimmerling toward Palestinians as one of "conceptual neglect." Kimmerling is a prominent critical Israeli scholar and the originator of the term *politicide*, which draws on work on *urbicide* from the 1960s and further helped to inspire the research on *cybercide* and *spaciocide* that is mentioned later in this chapter. Yet despite this critical import, Zureik (2001, 222) argues that Kimmerling neglects the agency of both Palestinians and American Indians in struggles over land due to Kimmerling's primary focus on the effects of the differing size of frontier territory in the United States (larger) and Palestine (smaller). This is notable because it indicates the challenge of reconfiguring discourses of rationality even among those who, like Kimmerling, actively and persuasively criticize the violence of the Israeli state.

The erasure of Palestinian agency also encourages misconceptions based in ideas of Western exceptionalism. Tal Golan (2004, iv) has suggested that the overwhelming support for science and technology in Israel enabled it, among other nations founded after World War II, to become "the *only* country that reached a per capita income level approaching that of the long-independent Western nations." Yet Golan implies that this wealth has also created an escapist academic culture that largely ignores the history of Israeli technoscience. Ongoing modernizing discourses exceed academic debate, however. Indeed, they are integral to land management practices that privilege maps. The broader recasting and reworking of the soil was in part an attempt to make it easier to map and thereby make the maps more legible for those reading them. Latour (1987, 254) has brought attention to the need for the modern geographies to continue to adapt to the map:

When we use a map, we rarely compare what is written on the map with the landscape—to be capable of such a feat you would need to be yourselves a well-trained topographer, that is, to be *closer* to the profession of geographer. No, we most often *compare* the readings on the map with the road *signs* written in the *same* language. The outside world is fit for an application of the map only when all its relevant features have themselves been written and marked by beacons, landmarks, boards, arrows, street names and so on.

Claiming that one would need to be a topographer in order to match map and land, rather than map and signs, Latour highlights the fact that the need for cartographers is almost as great when remaking the soil as it is when making a map.

In fact, cartographers were an intrinsic part of Israeli efforts to recondition material space, and they took part in furthering the modernist development discourses of geographers more broadly. Amiran, discussed in chapter 1, used Israeli land management methods to justify the Israeli hold on territory.[19] He characterizes the entire eleven-hundred-year period, from the seventh-century Arab conquest to the founding of modern Israel, as "centuries of decay when the land of Israel turned into an impoverished, malaria-stricken country, suffering both from maladministration and from a backward population." In contrast, Amiran (1987, 311) argues that Israeli developers created newfound wealth "with enterprise and modern agricultural techniques" in an era when "large-scale projects reclaimed the land which had deteriorated throughout centuries." He states, "Israel is an outstanding example illustrating that the same earth—the same natural resources—are put to different uses by different people, employing different technologies. Israel illustrates that land of great value to one population was entirely useless to another one" (ibid., 309). Amiran (ibid., 299–300) thereby implies that Palestinians had no right to Palestine, because they "had neither the means nor the initiative to cultivate it." Such assertions are directly tied to expropriation. Raja Shehadeh has contended that the spread of *natsh*, a local thistle gathered for making brooms and similar household implements, has been exploited by settlers in legal settings to expropriate territory since 1967. Paraphrasing their reasoning, Shehadeh (2007, 52–53) writes, "'The land is full of *natsh*. I saw it with my own eyes.' Meaning: What more proof could anyone want that the land was uncultivated and therefore public land that the Israeli settlers could use as their own?"

The practice of alleging mismanagement in order to seize land is not unique to Israel and Palestine. Instead, it is evocative of urban renewal practices that were dominant in Europe and North America over the course of the twentieth century (Carmon 1999), and that likewise built on colonial tropes.[20] The US Agency for International Development and others, for example, have made ongoing efforts to carve gaps into the Jerusalem hills to allow for the sloping, circuitous, and segregated highway for Palestinians, to complement and circumvent the more direct highways for Israelis. The latter are more than a little reminiscent of Rudolf Mrázek's (2002, 8) description of road-building efforts in Dutch Indonesia: "The newness, the hardness, the cleanliness—it was the roads' modernity."

The emphasis on cleanliness carried over to urban renewal efforts within Israel as well. Soon after Israeli troops marched into the Old City of Jerusalem in 1967, the Maghrebi Quarter, a historic Palestinian neighborhood that stood adjacent to the Western Wall in the Old City, was cleared. The act was represented as a "cleaning up process"—one that resulted in several deaths of residents who were crushed by bulldozers in the rush to demolish the historic neighborhood (Abowd 2000, 9). Planners viewed the houses, including a mosque from the Middle Ages (Hasson 2012b), as decaying and cluttered ruins (Ricca 2007, 6), and justified demolishing the centuries-old homes, which were fully inhabited, on the grounds that it was for "public use" (Abowd 2000)—meaning, the Israeli and international public. They thereby echoed the claims of *eminent domain* in the United States and problematic definition of a narrow public that it invokes. Eminent domain is the ability of the state to buy or take over areas to use them for the public good. In practice, eminent domain is often justified by arguing that the areas are "blighted" and not being put to adequate use. The implication, that land must be used in order for tenants to retain ownership, resembles the more recent contentions made in Israel. These international connections are far from surprising, given that many Israeli geographers and planners were trained in the United States in the early decades of the modern state of Israel (Carmon 1999).

In recent years, transformations in the landscape continue to be fully imbricated with attempts to frame Israelis as representatives of modernity in the region.[21] This dates to Theodor Herzl's iconic slogan, "A land without a people for a people without a land," which presented Palestine as empty territory in order to both recruit Jewish colonists and advocate for a Jewish state there. Indeed, throughout the twentieth century up to the present day, authorities have dammed streams, rerouted springs, and demolished ancient cisterns—tanks for collecting rainwater—in the name of a narrowly defined modernity that favored visibility, quantification, and clarity. Pine forests have been planted over the ruins of villages to override local scrub vegetation, allegedly making it more productive—that is, more like the situation in Europe (Cohen 1993, 1994, 2000, 2002). Large tracts, particularly in areas inhabited by Bedouins in the Negev, have been seized amid accusations that the Bedouin "failed" to develop the landscape.[22] This neglects how their long habitation apparently resulted in less damaging transformations than the enormous undertakings of the comparatively short history of modern Israel (Adalah 2013; Falah 1985, 1989; Human Rights Watch 2008; Shamir 1996). These efforts are both due to explicitly political motives and because traditional forms of use, such as the maintenance of olive groves and collection of wild plants, are not widely recognized as a type of agriculture. This even has resulted in a political opposition between olive and pine trees, with pine forests identified as "true" (read: European) nature (Braverman 2009).

In this context, once many Palestinian populations were forcibly removed, new exclusionary notions of agriculture and industry took hold. In time they largely replaced the intertwined forms of modern and indigenous knowledge by which the land was previously shaped and known. If those Palestinians who remained attempted to enter the areas claimed by the recently established government—for example, to pick herbs for cooking—they could be arrested and charged with trespass for reasons of security (Gurvitz 2011).[23] These movement restrictions in turn have enabled the demolition of Palestinian homes and seizure of agricultural areas in the West Bank on the grounds that the lands—even those that local Palestinians are prohibited from entering—are not being (properly) used.

Recent scholarship on infrastructure and occupation has drawn attention to the ways that forms of control are built into both social and material landscapes (Aouragh and Chakravartty 2016; Mitchell 2014; Salamanca 2014). Sharon Rotbard (2015) argues that Tel Aviv was built through the destruction and subordination of the nearby Palestinian city of Jaffa. Rotbard shows how both cities were shaped through discourses and practices that misleadingly cast Tel Aviv as modern, and Jaffa as its opposite. Zureik (2001, 214) has explored the role of surveillance in administrative infrastructure in the census. He refers in particular to the Kafkaesque category of the "present-absentee," and its continued use to disenfranchise tens of thousands of Palestinians who were pushed out of their homes during the 1948 war, but who nonetheless fled to areas that later became part of the state of Israel.[24] Such research provides an important corrective to a long legacy of work that neglects the effects of colonialism in development discourse and cartography.

These legacies of development and related regimes of international funding continue to shape the landscape in the present day. Indeed, Arthur Doerr, Jerome Coling, and William Kerr (1970, 337) have even argued that Israeli methods should be exported elsewhere, claiming that the lessons of the Israelis in agriculture "must be employed with swift and sure efficiency in the world's developing areas." Yet as with the earlier discourse that posits Israelis as modernizers, this recommendation ignores not only the ongoing military occupation but also the immense level of foreign funding that was granted to the Israeli state and private enterprise (Dattel 2013; Nitzan and Bichler 2002).

International aid was granted largely due to geopolitical considerations in the context of the Cold War (Hecht 2011), and the funding that Israel received was much higher and freer of restrictions than the support offered to other postcolonial states in the region. Yet although it was partly an attempt to atone for the horrors of the Holocaust, international funding to Israel also depended on the international perception that Israel was a modern state. European and North American countries sought to establish Israel as a Western foothold in the Middle East. This would allow them to have some proxy control in the region, as a partial replacement for the direct colonial control of the nineteenth and early twentieth centuries.[25] Yet the definition of Israelis as "allies" was also due to international power imbalances on the grounds of race and global region—namely, because Israelis were characterized as racially white and culturally Western.

Israeli and some Palestinian developers actively pursued this characterization, seeking to conform to and improve on Western development ideals. The consequences for Palestinians often have been dire, including the lasting and at times unexpected impacts of the sort explored in the rest of this book. Therefore, the discourses and practices of development were crucial for present as well as future attempts to manage specific areas. They are especially relevant in regions that are intensively developed and redeveloped (Hommels 2005) through the use of geographic data and digital maps.[26]

Landscapes and Labscapes in STS

An emphasis on the production of landscape, then, reveals how international modernization discourse in Palestine and Israel built on colonial methods of managing people and land. By highlighting the historical and geographical legacies of knowledge, such studies complement work in STS, which emphasizes the specific practices and materials that are used and reworked during the production of knowledge. Researchers notably have explored how the pristine laboratory, as a major arena of knowledge production, was only made possible by the exclusion of the rest of the landscape as the messy, unpredictable field (Latour 1988; Latour and Woolgar 1986). In classic STS ethnographies, however, the focus has frequently been on the laboratory alone, as scholars underscore how labs were created as unique spaces for experimentation (Karvonen and Van Heur 2014). To give just one iconic example out of many, Latour and Steve Woolgar (1986, 127, emphasis added) look at the constitution of scientific objects, while noting, "An object [of scientific study] is simply a signal distinct from the *background* of the field and the noise of the instruments." They thereby paraphrase a particular scientific worldview where the field becomes an insignificant backdrop of science.

Likewise, Karin Knorr Cetina (1981, 1999) has analyzed materiality, space, and broader patterns of knowledge production. Yet she stresses the ways that materiality is enacted within the laboratory, whereas broader patterns appear primarily in terms of social organization. Even so, her work simultaneously points to how the lab and field may be studied together. Given this, Knorr Cetina's (1999, 85) assertion that "tools and resources are the capital without which a lab could not maintain its place in the field," although intended to refer to a scientific field, can also productively be applied to the research field (in the sense of *fieldwork)* as well as the upkeep of related lands, buildings, and infrastructure. The focus on the laboratory at the expense of the field is partly a result of the privileging of laboratory sciences more broadly. Robert E. Kohler has investigated the impact of the spread of lab sciences over the course of the twentieth century. Kohler (2002) provides a detailed study of how field biologists responded to the increased attention to the laboratory and experimentation in the United States by developing notions of "nature's experiments" (ibid., 214).

In recent years, scholars have increasingly sought to blur and mediate dichotomies between the lab and field (Kohler 2002). For instance, studies of "urban laboratories" specifically challenge conceptions of the laboratory as a separate and sterile space (Karvonen and Van Heur 2014). This has led to more detailed attention to place making in scientific practice

(Gieryn 2000, 2006), and a broader interest in how the field itself is also produced. Likewise Pankaj Sekhsaria (2011, 2013) has explored how scientific instrument building relates to geographic contexts, while mobilizing material practices and knowledges that exceed narrow definitions of science. This book builds on and broadens these studies of the materiality of technology in both STS (Wyatt 2008a) and geographical research (Clark, Massey, and Sarre 2008).

Although there is less explicit focus on the production of space, scholars working in Palestine and Israel have extensively analyzed the relationship between ideology and the reshaping of the land. Meron Benvenisti points out the destructive consequences of the implicit valuing of rational knowledge in Israel. He notes how the ultimate destruction of Palestinian villages after 1948 was made possible precisely because those villages did not fit into the rational patterns that the Zionists expected of developed geographies. Benvenisti (2000, 56) states that to members of the Jewish communities, the Palestinian communities were "white patches—terra incognita. ... These towns, villages, and neighborhoods had no place in the Jews' perception of the homeland's landscape. They were just a formless, random collection of three-dimensional entities." In so doing, Benvenisti equates the towns' perceived lack of *form*—that is, of a pattern that would appear rational and organized to the Israelis—with their eventual near erasure. This erasure consists of the omission from both the map and, ultimately, landscape itself.[27]

Research on ideology is increasingly drawing on STS and geographic critiques of rationality and scientific knowledge to demonstrate the damaging effects of land management policy. Building on the literature on *politicide*, Nurhan Abujidi (2014) speaks of *urbicide* in Palestine, Miriyam Aouragh (2015) refers to the politics of technology under the occupation as *cybercide*, while Sari Hanafi (2009) calls the occupation's policies a form of *spaciocide*. Weizman (2002, 2007) refers to them as the *politics of verticality*, making the claim that the two-dimensional character of traditional paper maps and related thinking is insufficient to capture an occupation whose forms of segregation include aerial surveillance as well as road tunnels that stretch deeply underground. Nonetheless, the very use of *verticality* may have the unfortunate effect of implicitly reinforcing the primacy of Western rationality by framing discussions of landscape in Cartesian terms—that is, as being reducible to three separate numerical axes. This shows in particular the greater role for STS research in combining the study of space with analysis of the practices of fact making and observation.

To offer another example, in the geographer Denis Wood's recent discussion of "nonreactionary" maps—such as *The Atlas of Palestine, 1948* (Abu Sitta 2004) and the UN closure map (UNOCHA 2008)—he allows that facts can be selected according to specific ideologies. Wood (2010, 246–247) also notes that ideology is not the only contributing factor in a politics that is predicated on observation: "Each chosen fact is as factual as can be. None has invented data. None is a fantasy. You can go there … and check it out."[28] Yet despite Wood's slightly humorous appeal to commonsense efforts to "check it out," it is important to emphasize the different observations that particular individuals and groups may make based in part on their physical and social positioning in a specific landscape (see chapter 5). The study of precisely these practices of observation, of checking it out, are one of the many innovations of STS and related work.

Building on this existing research, it is necessary to pay closer attention to how subjectivities are materially situated. For it is not simply that "planning decisions are often made not according to criteria of economic sustainability, ecology or efficiency of services, but to serve strategic and national agendas" (Weizman 2002). In addition, sustainability, ecology, efficiency, and strategic agendas are *themselves* all fully enmeshed in political and geographical debates. In the next section, I introduce the literatures in critical GIS and STS more fully, and outline the potential of both for a more nuanced understanding of the links between past empires and development practices, on the one hand, and contemporary digital technologies, on the other.

GIS and Landscapes of Technological Practice

The strong interrelationships between land and map have altered, but not attenuated, with the onset of computer cartography. Indeed, far from signaling a break with cartographic methods grounded in empire, digital cartographers have built on colonial legacies. Empire and technological advancement were tightly coupled throughout the twentieth century. It is no accident that the oldest university in Israel, established in 1912 when the area was under Ottoman rule, is the Technion—the Israel Institute of Technology. In the early years of the Israeli state, technological knowledge helped to make possible the kinds of comprehensive transformations of the landscape just discussed. In time, wholesale modernization efforts, combined with international economic and military support, also enabled relative financial security. Yet although it required ongoing work and determination, Israel's financial affluence did not develop out of some natural progression. Instead, as mentioned above, it was related to broader military goals on an international level as well as Israel's display of adherence to alleged European values, including the championing of science and technology.

As a result of both the high level of international funding and tremendous efforts at modernization within Israel, Tel Aviv has fast become a "Silicon Wadi" (Gordon 2011, 157)—a counterpart to California's Silicon Valley.[29] Israel presently hosts branches of numerous companies such as Intel, Motorola, IBM, Microsoft, and HP, to name only a few (Rashi Foundation 2012).[30] As in much of Europe and North America, the technology developed in Israel has often been first designed and used for military purposes before being adapted for consumers. There is ongoing cooperation between universities and the military in high-tech research—a relationship that has been "widely discussed" in Israel (Sheizaf 2013). Despite the links between the military and geographic technology (Livingstone 1992), however, the connections between colonialism and specific technical details are only beginning to be understood.

It is not surprising then that, as Cindi Katz has observed, those in power are frequently infatuated with maps. Katz (2001, 1215) refers to it as going "map-crazy" and hypothesizes that it was "no more than a form of 'gun-crazy' once removed." This interweaving of development and digital cartography begs further investigation. It can be achieved by combining literature on cartography and technology—work that is influenced by STS, with its focus on how broader social contexts affect the inner workings of technologies—with research in critical GIS, an emerging theme in geography.

The vast majority of the labor that goes into making a map is not readily visible in the images that are finalized and printed from it. With this in mind, both critical GIS and STS emphasize the ways that cartography is an extended process, from the initial collecting of data, to choices of how data are categorized and displayed, and on through the map's ultimate dissemination and use. This contrasts with the geographic scholarship on cartography and empire as explored earlier in this chapter, which has traditionally centered more on the contents and political effects of maps once they are completed (e.g., Harley 1989). Critical GIS and STS thus enable a more textured understanding of specific practices. As such, they can help to link studies of empire with research that reveals how contemporary practices continue to draw on the problematic development and modernizing discourses explored in the previous section. To better understand the full process of making a map, I next describe the use of GIS cartography. I then turn to a more detailed examination of the relevant critical GIS and STS literatures.

Maps as Interactive Visual Databases

To understand the relationships between contemporary cartography and empire, it is necessary to first say more about how a digital mapmaking program works. Far more complex than a simple image viewer, GIS is a way of visualizing and analyzing multiple databases at once. Although there are many different interpretations of the goals of GIS, the adoption of GIS has coincided with an increasing emphasis on statistics and quantitative analysis in cartography (Longley and Batty 2003). Esri's ArcGIS Desktop is currently one of the most widespread GIS software packages in use. Similar to a blank Word document, a new GIS map document often simply consists of a white window on the screen. After creating it, a user first adds layers to the map. So there might be one layer for roads, another for rivers, and a third for buildings. These are stored as separate files and are added in turn to the current document. They are then viewed and modified together to make a final map.

The kinds of layers that are loaded into map files are themselves data sets that can be linked to tables with a large number of related attributes, such as length, density, and textual labels, among others. The result is that the map becomes interactive. Users not only see the location of a dot indicating a city, for instance, but by clicking on the dot with the mouse, they can bring up information for that city, including its name, population, and so on—provided that the map has been designed accordingly. But the potential to bring information tables together, though large, is not infinite. Large data sets might cause the program to crash unexpectedly, or can lead to long processing times.

Once the desired layers are visible and the map layout is set, printing or saving the file as an image becomes almost an afterthought. As Valérie November, Eduardo Camacho-Hübner, and Latour (2010, 584) point out, rather than emphasizing the printed map, digital cartography has "*rematerialized* the whole chain of production"—a chain that "requires people, skills, energy, software, and institutions" that all contribute to the "constantly changing quality of the data." The interactive yet ephemeral GIS maps on a computer screen can be produced by a multitude of cartographers. They stretch from the user to the graphic designer, to the one who manages the database, and on back to the initial fieldworker with a GPS unit.

The GIS map file can be continually updated as data circulates and new maps are made, in ways that though they draw on older practices, nonetheless stand in stark contrast to the apparent fixity of paper maps. Paper maps too had their legions of support teams, but they tended to focus much more on a finished product—a process that differs somewhat from the constant feedback and recirculation involved in making digital maps (November, Camacho-Hübner, and Latour 2010). Once a paper map was completed, it could be difficult to modify or update, and it was far easier to explicitly add new features by drawing on the map than it was to remove existing features. For this reason, the features on paper maps tended to be fixed in place, and the updates were clearly visible as additional fixed features. In contrast, GIS maps are much more susceptible to alteration as circumstances change. Features can be removed as easily as they can be added. GIS mapmakers tend to emphasize the printed product far less, and their maps are more susceptible to process and change. In spite of this altered emphasis, digital cartography also exhibits strong continuities with paper mapmaking, as I explore in the next section on critical GIS.

Critical GIS: The Work of Making Maps

Just as the relationship between maps and empire has been critiqued, scrutiny has been cast on the role of GIS maps in mechanisms of control. In particular, scholars have pointed to continuities between digital cartography and previous cartographic methods. If the introduction of computer maps represented an increase in malleability, it has also reinforced some of the fixities associated with paper maps. Chief among these is the privileging of particular forms of mathematics and quantification. In a move that is far from unknown in other disciplines, the increasing importance of computers has made it ever more attractive for cartographers to quantify every aspect of the land and then analyze it solely in terms of tables of attributes of the sort described above. Maps of landscapes could well include elements like memory, sensory impressions, stories, and extensive textual description. More and more, however, official mapmaking has been reduced to a process of counting trees, people, buildings, and other visible objects, and creating tables of data. The results both derive from and reinforce a narrow conception of the types of experience and observation that are fit to be called knowledge.

This particular, narrow use of GIS has also been subject to critique from its earliest history. In direct opposition to the sometimes-euphoric reactions of GIS users, human geographers' assessments of GIS in the 1990s consisted of strongly worded condemnations of a technology that was viewed as taking over the discipline of geography. This led to sometimes-acrimonious debates that Peter Fisher and Donald Unwin (2005, 3) have referred to as the "GI wars," paralleling the "science wars" of roughly the same period. Particularly notable are the extensive critiques of Pickles's (1995) edited book, *Ground Truth*. In the context of these debates, Peter Taylor (1990) famously referred to GIS as a new form of imperialism (see also Fisher and Unwin 2005, 1).

Although the concerns of critical scholars are justified, the reactions against GIS and positivism in general have gone so far as to characterize all quantitative methods as clear, monolithic, and unchanging (Wyly 2011). In addition, many of the early criticisms were made by those without experience in using GIS technology—partly because it was new at the time—and they of necessity focused more on the role of GIS in society, to the exclusion of the different ways that GIS itself might be changed and used. In the ensuing decades, newer generations of scholars with expertise in both GIS and critical theory have developed the subdiscipline of critical GIS, which is associated with both critical cartography and GIS analysis (Crampton and Krygier 2006).

Critical GIS researchers, in keeping with other critical cartographers, argue that even the most scientific or natural maps are subjective in specific visual and textual ways. For all forms of cartographic knowledge are conditioned by historical and geographical systems of power (Dodge, Kitchin, and Perkins 2009; Massey 1994; Piper 2002). As a result, maps never simply convey information in a direct and unmediated manner, but instead they are invested with the ability to incorporate some forms of information while omitting others (Dodge, Kitchin, and Perkins 2009).[31] In short, maps lie and tell half-truths. This is the reason that one broadly influential book on cartography is titled *How to Lie with Maps* (Monmonier 1991).

There is widespread scholarly agreement that maps can lie, and many would agree that all maps do so, if only out of necessity. The contentious question remains whether a map can ever tell the truth, or if factual truth is even the most important goal. For the notion of *truth* implies that there is an alignment between a cartographer's intentions, the resulting map, and the ways others interpret the map. Furthermore, in the cartographic literature there is a tension between an emphasis on distortions that are believed to be deliberate or intentional—such as adding trees around a map of a proposed development to make it seem more appealing (ibid.)—and those that are seen as nondeliberate or even inevitable—such as the distortions that arise from flattening or projecting a three-dimensional sphere, such as a globe, onto a two-dimensional plane, like that of many maps. Thus, Timothy Wallace and Charles van den Heuvel (2005) productively explore the varying notions used to specify the truth of the information in maps, many of them normative, such as *error, quality, accuracy*, and *precision*.

Yet although they remain a crucial area of inquiry, intentions do not tell the whole story. J. B. Harley (1989, 13) has remarked that power was naturalized to such an extent in cartography that it was not always exercised consciously. For this reason, he calls the map the "silent arbiter of power." This reflects the lies that cartographers tell themselves—or the lies they don't know they're telling—as much as the lies, intentional or otherwise, that they tell others. Although it is sometimes discussed in critical GIS research, the nondeliberate shaping of the truth is less well understood in the cartographic literature. It is more extensive than is typically acknowledged, and also can coincide with the act of deliberately moving a border or erasing a village.

For these reasons, this book is not concerned primarily with the conscious–unconscious distinction, or an intellectual history of the cognitive, personal, or psychological motives of individual cartographers.[32] Indeed, if privilege is perhaps more evident to those who don't have much of it, then the question of whether some forms of science *deliberately* and *consciously* perpetuate social injustice is perhaps less interesting than how—regardless of intentions, or even with the best of intentions and actions—those injustices are nonetheless perpetuated. While I do not shy away from calling out deliberate injustice, I also emphasize the insidious ways that power imbalances can serve to circumscribe personal intentions. This sustains a process of perpetuation that is present not only in interpersonal relationships and sovereign governance, two notable examples, but also in even the most mathematical technical details of the most digital technologies.

Attention to technical practices are also key, because there is a danger that a critique of quantitative methods might end up only reversing the imbalance by simply privileging qualitative over quantitative research. Yet given the stringent early criticisms of GIS, many scholars of maps and empire understandably have hesitated to make maps themselves—namely, to take part in the practices that they so effectively critique.

In one prominent example, Mitchell (2002) argues for a better understanding of the nuances of quantitative methods in particular contexts. Still, he stops short of proposing countercartographies, or allowing such efforts to influence his own methods. Given mapmaking's colonial histories, some researchers are reluctant to allow that it might ever be possible to make de- or anticolonial maps. This contrasts with the positions of critical GIS scholars, who are frequently cartographers as well as social theorists. Certainly, textual critique remains incredibly relevant. As a result of its interdisciplinary position, however, critical GIS has a unique vantage point in terms of being reflexive with respect to the "rift" between critical social theory and (primarily quantitative) spatial analysis (Kwan 2004). Addressing this rift is central to expanding the reflexivity of knowledge.

Thus, critical GIS scholars argue against privileging textual critique, which has long been favored in critical theory, over other methods like critical visualization or quantitative analysis.[33] They are also cautious, though, because quantitative knowledge is invested with particular power, and wider public recognition of the value of social theory and the humanities is sorely needed. So it pays to be wary of the tendency to overvalue GIS in ways that can marginalize other geographic methods (St. Martin and Wing 2007). Nevertheless, the focus on how different methods are interrelated in turn allows for greater nuance in accounts of the historically and culturally specific division of academic methods and disciplines. This also includes, for instance, the politically loaded and at times unhelpful separation of different spheres. To give just one example, the professional study of GIS is organized on par with graphic design and related fields like urban planning. The equally relevant fields of information technology and server management tend to be concentrated separately, in trade schools. The divisions between quantitative and qualitative knowledge, and those between disciplines, reflect the idiosyncratic histories of knowledge and universities, yet they are essential to understanding the social role of technologies like GIS.

In light of these assessments, critical GIS makes room for a critique of the separation between math and text, and terms like *qualitative* and *quantitative*, while extending the important examination of method into the critical methods of the scholars themselves. Marianna Pavlovskaya (2006, 2004) convincingly maintains that the apparent divide between quantitative and qualitative research is actually a "continuum," rendering the two fully compatible in spatial analysis using GIS, which she characterizes as "neither a quantitative nor qualitative tool." In a similar spirit, work in critical GIS more broadly has furthered the wider understanding that empirical maps are only one of the myriad ways of displaying data—ways that include schematic visualizations, artistic representations, and maps of embodied geographic knowledge. This has also helped scholars to understand how, rather than making older technologies wholly obsolete, GIS simultaneously incorporates a variety of technologies, such as pen and paper, GPS, and surveying equipment, to name just a few (Crampton and Krygier 2006; Pavlovskaya 2006).

However, this broader focus of the academic literature has not yet filtered into popular knowledge. Indeed, the data-focused capabilities of GIS appear to have the potential to reduce all spatial questions to mathematical models and formulas, so it is perhaps not surprising that much critical GIS focuses on the role of quantification in geography. Drawing on current research in geography and postcolonial studies, Joanne P. Sharp (2009, 64) has noted the colonial roots of the favoring of quantification in GIS, arguing that European colonial regimes privileged space, including "making space mathematical and ordered (challenging the indigenous ordering of space) in such a way as to render the colony most efficiently known and governable." She thus draws links between the colonial cartographic legacies investigated earlier and the digital maps that have proliferated in recent years.

Critical GIS scholars also have been quick to point out that GIS is not static. Any data table used in connection with a GIS map might equally contain numbers or letters. This means that at the very least, there is greater qualitative and descriptive capacity than first imagined, despite the focus on classification. In addition, reconfigurations can occur on many levels, from alterations of hardware and software—such as the development of online cartographic services like Ushahidi, Mapbox, and CartoDB—to the creative use of existing techniques like fuzzy boundaries (e.g., Ahlqvist 2005), to the expansion of communities of cartographers, both in person, including through participatory GIS (Scholten, van de Velde, and van Manen 2009, 3) and by integrating social media with services like Google Earth (Farman 2010). These methods by no means guarantee that the maps will be somehow revolutionary or socially just, but they do increase the opportunities to discuss and realize the varied potential of maps. Thus Michael Curry's (1998) early emphasis on the limitations of GIS, as a technology that underscores scientific values such as universalism, disinterestedness, and skepticism, may become less relevant over time. To date, however, it is difficult to tell, and a more detailed history of the social development of GIS as a technology, as opposed to digital cartography more generally, is lacking (O'Sullivan 2006).

Nonetheless, since its advent, one major reconfiguration of digital cartography is apparent. Computer mapmaking has seen the relative democratization of cartography as selections of professional data sets and software become available online. GIS techniques have begun to be combined with more open data and tools (Elwood 2009; Farman 2010), thereby being incorporated into practices that stretch beyond professional cartography's traditional focus (Scholten, van de Velde, and van Manen 2009). At the same time, GIS software is far from inexpensive, and many sophisticated data sets remain under the exclusive control of national governments and private corporations. This includes everything from updated higher-resolution aerial photographs to data of underground infrastructure like plumbing pipes and information and communications technology cables (Dodge, Kitchin, and Perkins 2009).

Even so, the open source movement is gaining currency in cartography, allowing for the wider use of geographic technology. Open Street Map, a Google Maps alternative, is available for public editing, and QGIS is currently available without charge. Alternative cartographic methods on the whole are also expanding in ways that often involve transnational social networks. One example includes Engineers without Borders Palestine's (n.d.) efforts, in concert with JumpStart International, to map West Bank roads simply by driving along them and recording the route on personal phones. Engineers without Borders also explores alternative engineering more broadly, including making solar coffeemakers (repurposed television satellite dishes) for Bedouin families in unrecognized villages, and developing mobile homes that can be moved at a moment's notice for Palestinian communities under threat of demolition (interview 12, Palestinian civil engineer). Another related project is the ongoing effort to map Israeli protest marches from above, to demonstrate the extent of internal dissent, which might be downplayed in mainstream news outlets. As part of a growing transnational movement in alternative cartography, cartographers tape digital cameras to kites and helium balloons, and this allows them to create their own aerial photos at minimal expense (interview 21, freelance Israeli NGO cartographer). Unfortunately, many such projects are short lived, and subject to even greater issues of international politics, funding pressures, institutional challenges, and enforced precarity than those that shape more formal cartographic projects.

Although concerned with the politics of maps, critical GIS is only beginning to incorporate and build on postcolonial theory, and the case studies still often focus on Europe and North America. Even so, in the global South there have been a number of participatory efforts—more formally known as public participation GIS—albeit frequently in cooperation with universities from Europe, Australia, and North America. These include projects to incorporate indigenous groups in making maps of the spaces where they live (e.g., InfoAmazonia 2013) in order to attenuate exclusionary mapping in the construction of conservation areas and national parks (Harris and Hazen 2006).

Yet there are concerns with participatory mapping as well. To the extent that cartography represents a dominant form of knowledge, further incorporating indigenous groups into dominant knowledge systems can pigeonhole them. For example, once any particular group is "given" land on a map, they may have difficulties accessing places outside that official area. Likewise, the expansion of GIS internationally can lead to the entrenchment of technocratic elites who only reinforce existing power hierarchies (Dunne et al. 1999), and the resilience of state mapping organizations can steamroll alternative cartographic efforts, no matter how successful (Duncan 2006; Radcliffe 2009). Alternative GIS has yet to fully incorporate the incredible variety of everyday spatial practices. Public participation GIS scholars, however, continue to innovate in terms of how indigenous conceptions of space might further inform academic theory and GIS practice (Moore 2007).[34] Given that GIS technology is itself situated within asymmetrical relationships and legacies of power, then perhaps paradoxically, a symmetrical framework is a uniquely effective way to study such asymmetrical legacies. With this in mind, I now turn to an analysis of the related conception of *symmetry*, which is a central focus of STS research.

Asymmetrical Violence and the Violence of Symmetry

Although STS and critical GIS are distinct, the two share similar concerns (Schuurman 1999, 2000) and methods that allow for an examination of the relationships between empire, maps, and digital technology. Chief among the latter is the notion of *symmetry* in STS. As noted in chapter 1, symmetry involves artificially constructing a comparison among diverse entities in order to highlight differences in their very constitution. So while geography provides STS with detailed studies of space, to geography STS contributes the notion of *symmetry* as well as a concern for the details of technoscientific practice (Coutard and Guy 2007).

Yet before further analyzing *symmetry*, it is important to first situate it within the unique context of STS—a relatively new field with its roots in the sociology of scientific knowledge of the 1960s and 1970s. Although increasingly organized as a discipline, STS is thoroughly interdisciplinary, with its practitioners united primarily in terms of the broad contours of methodology along with a focus on contemporary or recent issues in society, science, and technology as well as a critical theoretical approach. In terms of that approach, STS works in tandem with research in sociology and anthropology while drawing on, but remaining distinct from, the history and philosophy of science.

In this, the founding innovation of STS research is that society is not a mere add-on to the history of science and technology. Instead, social, political, economic, and cultural concerns shape the very *content* of technoscience. With a focus on the role of controversy and contestation in the production of knowledge, STS research has moved away from "lone genius" histories of invention, and toward an understanding of how knowledge is situated within social and material contexts at multiple scales. The Internet is a prime example because its standardization and hierarchical architectures have their origins in UK and US military research, but its protocols were influenced by international networks of independent researchers operating according to liberal ideals of the free exchange of knowledge. STS research often revolves around contemporary or recent concerns, and includes ethnographic fieldwork with focused case studies that allow for an examination of the relationships between social contexts and technoscientific details. As an interdisciplinary field that combines critical theory with empirical and analytical concerns, STS thus provides a nuanced and singular perspective on histories and practices of knowledge making.

At its formation, STS continued the legacy of the Strong Programme (Barnes, Bloor, and Henry 1996, 19; Bloor 1976). Researchers investigated the ways in which specific forms of technology are culturally and historically contingent (Bijker 1995), while drawing increasing attention to the importance of qualitative, ethnographic methods in the social study of science and technology. This has led to a body of literature that tends to concentrate on local technological practice, although scholars are increasingly moving to link case study research across broader political, economic, and social systems (Bank and Van Heur 2007; Wyatt and Balmer 2007). Related research has also considered the influence of economics and politics on the development of professional networks (MacKenzie and Wajcman 1999; Van Heur 2009), and the "boundary work" that is done to demarcate between areas of knowledge (Gieryn 1999; Halffman 2003; Henke and Gieryn 2008; Jasanoff 2004).

While research on boundary work in particular centers on the formation of divisions between disciplines as well as between experts and policy makers, Willem Halffman (2003, 57) has argued that "the resilience *and* dynamic of boundaries of science cannot be found in language alone." Halffman (ibid., 71) contends that "boundaries can even be embedded in the structure of an organization." This points to the need for a more substantial critique of landscapes of science, like that undertaken here, as it implies that a conception of science as a seamless enterprise, only occasionally marred by boundaries, does not deal seriously enough with the impact of boundaries on scientific practice.

Building on Halffman's simultaneous analysis of discourse and materialization, I take up Christopher Henke and Thomas Gieryn's (2008, 359) related call to research how "legitimate knowledge requires legitimizing places." For Henke and Gieryn, and prominent geographers like Doreen Massey (2005), such research includes the study of the spatial distribution of the branches within an institution, like a university or corporation, and between institutions. For the spatial organization of institutions both reflects and affects the constitution of knowledge. However, a fuller examination of the geographic distribution of science also requires an exploration of power, including the role of imperialism in manifesting particular knowledge practices. As such, by way of focusing my arguments for the importance of landscapes, I follow Anne Beaulieu, Sarah de Rijcke, and Bas Van Heur (2013, 29, 26), who emphasize the interaction between institutions and broader infrastructure in knowledge production, especially in terms of "how new actors come to be involved," "how empirical material is legitimized," and "how knowledge claims are validated in relation to existing and emergent forms of order." A consideration of how these processes are situated in geographic landscapes can only enrich the resulting analysis.

Cartography has not been considered at length in the STS literature. Latour has asserted that maps circulated as standardized knowledge tools or *immutable mobiles* that enabled decision making across vast distances. Indeed, the better part of Latour's (2005) iconic *Reassembling the Social* draws on mapping metaphors, and his current work highlights mapping controversies. Yet the cartographic "black box" is only occasionally opened in STS research—that is, with several notable exceptions, such as works by Ariel Handel (2009), Christine Leuenberger (2012), Leuenberger and Izhak Schnell (2010), and James C. Scott (1998). Given the concern in STS for the local specificity of technology, the need for a study of digital cartography is especially pressing. Latour claims (2005, 163) that "depending on which tracer we decide to follow [through society] we will embark on very different sorts of travels." With this in mind, the time has come to explore how, depending on *who* one is and *where* one travels, we end up with very different tracers.[35]

Symmetry developed as a response to mainstream approaches to the history of technoscience. In a form of circular reasoning, ideas and technologies were often said to be successful because they were useful, or simply "worked." Innumerable reasons might be given for why this was so, from the perseverance of individual investors, to political change, to the inherent durability or flexibility of a particular material. In contrast, it was argued that failed technologies simply were not useful, or didn't work—and no further explanation was necessary. This reasoning is at once tautological and asymmetrical. It's tautological because it claims that an artifact, like a particular GIS software package, succeeds because it's useful, and we know it's useful because it succeeded. It's asymmetrical because a successful technology might merit hundreds of pages, while an unsuccessful one merits only a brief mention.

Trevor Pinch and Wiebe Bijker's (2012) influential work serves as a corrective for this reasoning. For Pinch and Bijker, a *symmetrical* account is one that equally outlines the reasons for a technology's success and failure. They analyze the history of the bicycle, pointing out that it is not enough to say that the bicycle, in its current form, succeeded because it was safer and more reliable. Similarly, it isn't sufficient to note that other kinds of bikes failed because they simply weren't useful—without further explanation. Pinch and Bijker instead study the main types of bicycles that succeeded or failed at different moments in history, attempting to treat success and failure symmetrically by providing equivalent attention, and using equivalent reasoning for each. In their case, the use of symmetry allows them to show the problems with standard accounts, while also analyzing the heterogeneity of the causes of success and failure. Putting the reasons for success and failure side by side, symmetrically, allows them to demonstrate how unequally successful and failed bikes had been treated in the literature up to that point.

As a constructed measure for examining the complexity of power imbalances, symmetry can be incredibly useful for incorporating reflexivity and intersectionality, in terms of the intersections of different social identities and relations, into broader research design. Rather than a simple comparison, the goal of symmetry is to juxtapose subjects in order to analyze the very context and suitability of comparison and commensuration. As such, it is especially helpful in attempts to avoid reverting to dominant narratives in subjects where particular types of accounts have traditionally been alternately ignored, obviated, or stifled. Sally Wyatt (2008b) explores how symmetry was soon expanded to include alternative configurations, perhaps most famously in Actor-Network Theory attempts to be symmetrical with respect to an actor being human or nonhuman (Callon and Latour 1992). The concern with nonhuman actors emerged in part to better incorporate materiality into accounts influenced by social constructivism, which were sometimes accused of neglecting it. After all, a faulty part might cause a bike to fail, and that part then serves as one actor, albeit a nonhuman one, in the broader web of social and material groups that have invented, designed, built, and operated that particular bicycle.

Drawing on these prior applications of symmetry, this book sets up a symmetrical account between Palestinian and Israeli forms of cartography, as noted in chapter 1. The goal is to avoid the peril of the most common, all-too-simplistic account of development in Palestine and Israel. That account is eerily similar to the circular view of the history of technology that Pinch and Bijker critiqued, and embodies similar strands of superiority and inferiority. The simplistic tale runs something like this: Observers might note that "Israeli science and technology have succeeded because they are *better*"—and for "better" they might insert any number of claims of Israeli superiority that have prejudicial undertones, such as more European, racially white, rational, developed, democratic, organized, and so on. In contrast, observers claim that Palestinian technoscience has failed (which is not the case) because it is *worse*—worse because they are practiced by Palestinians who, it is (mistakenly) implied, are inherently inferior. In attempting to get out of this Eurocentric and even racist formulation of knowledge production, symmetry allows us to outline and de-emphasize such prejudicial accounts. Similar depictions crop up persistently even among the most attentive and dedicated scholars, given that it is so difficult to move beyond academic traditions that despite their many benefits, continue to be prejudiced after hundreds of years of colonialism, slavery, and imperialism.

So rather than attributing success or failure solely to motivations internal to an individual actor, symmetry allows for an assessment of patterns of imbalance within forms of knowledge. Nonetheless, it is also crucial to stress that it is a strategy, a thought experiment of sorts, whose aim is to highlight pervasive forms of erasure. When dealing with economic, social, and political injustice, symmetry has its own pitfalls. Most important, the symmetrical incorporation of an alternative account, by placing it alongside a dominant one, still involves defining that account in ways that have their own erasures. Thus symmetry is a form of commensuration. Even defining it as an *account* can be incredibly problematic, especially in cases where one dominant view stands in opposition to a plethora of other options. But the constructed nature of a symmetrical axis means that there is no single right way to propose a symmetry. Given this, I do not posit that a symmetry between Palestinian or Israeli science is perforce the only view. Other relevant symmetries might be posited in innumerable other ways. Yet in the current context of the occupation, which imposes a rigid divide between Palestinians and Israelis, the proposed symmetry is especially salient for exposing the injustices of that unequal divide and its very rigidity.

In order to make symmetry useful for research concerned with social justice, it is necessary to further explore its promises and dangers, in addition to the thematic connections between STS and critical and postcolonial theory. In STS, for example, Wyatt (1998, 2008b) actively incorporates concepts introduced by actors themselves. She pulls the notion of symmetry into a consideration of speech and authority that resonates with anthropological research. Postcolonial theory can help to further contextualize the politics and cultures of knowledge. For STS and geographic research that is concerned explicitly with social injustice often focus on the dangers inherent when particular voices go unheard. In contrast, Gayatri Spivak (1988, 1999) argues, for instance, that power imbalances affect the very constitution of voices themselves as well as the culturally specific focus on the notion of a *voice* (1988, 1999). It therefore is not enough to provide a "voice for the voiceless" when the very idea of a voice may itself stem from oppression and even work to silence other modes of (nonverbal) communication.

Spivak's analysis draws on one of Jacques Derrida's (1981) central arguments: power imbalances are far more insidious than just simple imbalances between two otherwise-equal sides. Instead, the constitution of each side is ontologically unequal. For example, one side may be created as a produced whole—a clearly delineated discourse or object—while the opposing side is not a whole but rather a dissipated nothingness. Although he is referring to discursive hierarchies, Derrida's claim is prescient for the study of the occupation. He outlines an opposition between one group that is recognized, defined, and dominant, and a second group that is dissipated, invisible or partially visible, and inconsistently defined, if even mentioned as a group in dominant discourse. Such oppositions are not "peaceful coexistence" but instead "violent hierarchy": "one of the two terms governs the other (axiologically, logically, etc.), or has the upper hand" (ibid., 41).

As a form of strategic play that was formulated independently from Derrida's critique, symmetry shares with deconstruction an attempt to account for the negation of particular groups. It does so by actively incorporating formerly disregarded experiences in the analysis. Hence, in the mid- to late twentieth century, it was relevant to speak of "Israeli science," yet Palestinian scientists might be ignored or disregarded as not "true" science—or if they were successful, they could be deemed as becoming scientists *despite* being Palestinian. So not only were the two groups in an unequal hierarchy; Israeli science also was an internationally recognized field with academic departments, libraries, and archives. In contrast, Palestinian institutions were alternately ignored and disparaged—and physically erased at times, either by using financial and political means to prevent their original founding, or else by physically demolishing them after the fact.

In this way, symmetry presents a means of critical reflection. It is an opportunity to better understand what happens when social groups are placed into rigid and unequal relationships like a military occupation. Symmetry is a strategic opportunity for critique rather than an end goal in itself. In the chapters that follow, I analyze Palestinian and Israeli cartography in order to better understand the process by which they are being constructed as the separate disciplines of "Palestinian" and "Israeli" cartography. I highlight the ways that the unjust landscapes of occupation differently shape the practices of those who call themselves Palestinian and/or Israeli cartographers. But as noted in chapter 1, this is not to suggest that this division between Palestinians and Israelis is either the only or most socially just way to practice cartography. Nonetheless, the maps throughout this book are evidence of how productive both Palestinian cartography and Israeli cartography are, and how innovative forms of resistance can be, despite the ongoing challenges.

Conclusion: Postcolonial GIS

Throughout this chapter I have argued for combining broader geographic literature on cartography and empire with postcolonial theory, as well as research in STS and critical GIS that investigates the specific materialities of technology. Such interdisciplinary practice furthers the move toward a postcolonial formulation of GIS, in keeping with a fuller awareness of the legacies and materialities of GIS technologies and practices. At the same time, it provides an opportunity to expand literature on colonialism and technology by incorporating considerations of how materiality and space influence digital knowledge, including even its most technical details. This shaping helps crystallize forms of control within particular scales at the expense of others. A symmetrical account thus allows for the study of how some scales, regions, and groups are produced—especially as *incompatible* or *inequivalent* groups—and others are aggressively erased.

The ensuing chapters represent a symmetrical analysis of the ways that geographic landscapes shape empirical cartography in Palestine and Israel. On the one hand, ArcGIS is a software package like many others. It is a sometimes-clunky way of combining tables of data and graphing them in many different colors. It is a means of compiling, within a single screen, the vast amounts of information collected at different resolutions and under varying conditions. As a cartographic technology with colonial roots, GIS allows representations of land to be seemingly divorced from the landscapes in which they were created, although they continue to be fully imbricated within them. Yet ArcGIS is also an advanced multidimensional interface. It permits ever more detailed studies of the intricate facets of the surface of the earth. It does so in ways that also combine existing methods and technological components—components that nevertheless are innovative in combination. Many aspects of these methods overlap with older forms of paper mapmaking, and stretch back to the heyday of colonialist appropriation of territory through maps. Thus GIS does not offer an escape or even corrective to past histories of domination, but likewise neither is it entirely fixed in its present form. Rather than resolving a singular, universal picture of the world, GIS enables ever more detailed disputes. In the next chapter, I turn to the early history and prehistory of GIS. I examine disputes over how to count populations in Palestine and Israel, and demonstrate that even abstract statistical maps are thoroughly influenced by the multiplicitous landscapes where they are made.

3 Removing Borders, Erasing Palestinians: Israeli Population Maps after 1967

Making these experiments sometimes made me feel like a kindergarten boy playing with colored papers, and afforded boundless amusement for my grandchildren.
—Roberto Bachi, *Graphical Rational Patterns*

It was decided to carry out the enumeration from house to house under curfew. ... The enumerator marked the doors of the houses enumerated with chalk to ensure an orderly and complete coverage.
—CBS, *West Bank of the Jordan, Gaza Strip, and Northern Sinai, Golan Heights*

The Methodology of Curfew

The quotations above serve to illustrate the intimate ties between population control and the development of statistical cartography. They date from the beginning of the Israeli occupation of the West Bank and Gaza Strip in 1967. The population census described in the second quote began only a little over two months after Jordanian and Israeli troops had ceased fighting street by street through the Old City of Jerusalem. The author of the first quotation, Bachi, was the head of the Israeli CBS, which was in charge of the enumeration, and he was at the height of his career. Bachi remained a key architect of the Israeli census for over twenty years, and was praised in diverse corners of the international scientific community for his innovative work in statistical cartography and geostatistics (e.g., "In Memory of Roberto Bachi" 1996). The quotation captures Bachi's romantic depiction of his academic research into his method of Graphical Rational Patterns (GRP), which were a new set of symbols for depicting numbers in a precise way on maps (see figure 3.1). The "colored papers" he refers to were transparent stickers or transfers that were applied when making a map, to

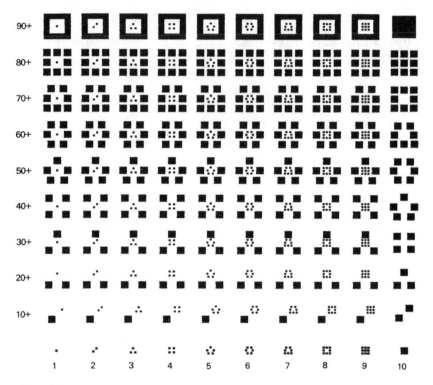

Figure 3.1
An example of Bachi's (1989, 339) GRPs: A chart of the symbols for the numbers from one to one hundred. Reprinted with permission.

prevent the cartographer from having to draw every symbol by hand. The second quotation comes from the official report of the 1967 census of the Palestinian Territories, carried out by the CBS that Bachi both founded and directed.[1] It describes the curfew imposed on the Palestinians, who were forced into their homes, or the home they happened to be closest to at the moment, their doors and walls marked with chalk, so that the count might be considered "orderly and complete."

There is a clear symbiosis between the colored papers and curfew. In the first instance, Bachi depicts himself cutting into the colored sheets, using glue to stick them into place on maps. In the second, Bachi is at the head of an army of enumerators, supported by the actual military (CBS 1967, iii), which physically marches through the streets, metaphorically cutting up groups of people and sending them back inside, fixing them in place, marking them with chalk in order to obtain a clear picture. This confluence of seemingly innocuous snips of paper, on the one hand, and the violence of putting entire regions under house arrest, on the other hand, gives an idea of why the national census, or systematic population count, resonates across such a wide array of research areas. The census encapsulates so many major themes of contemporary social theory: the falsity of the boundary between politics and technoscience; the creep of big data and minute slicing up of individuals and communities into predetermined check boxes like "age" and "nationality."

In many countries, the census was one of the main points of interaction between the masses and governments that sought to both discipline and depict them in ways that made sense to bureaucrats. In this context, Bachi's seeming glibness about his paper cutouts provides a cautionary note for studies of the census that do not take into account the ordinary and extraordinary coercions of quantification. For the violence of a census under military occupation is not so far distant from the context of the everyday work of national censuses more broadly. This chapter explores these connections between classification and sorting in statistics along with their forceful implications and applications on the ground. In the process, it shows computers in a new light, offering insight into the apparently seamless digital maps that nonetheless expertly, if imperfectly, conceal their connections to occupation and war. It therefore analyzes the ways that Israeli population cartography incorporated Palestinians through the very act of erasing them.

The relationship between state control and academic research has often been conceived in terms of the interplay of representation and materiality in shaping the landscape (Harvey 2005; Low and Lawrence-Zúñiga 2003; Smith 2008). In practice, however, the focus has been primarily on representation (e.g., Mitchell 1991).[2] In this chapter, instead of emphasizing how representations of space can shape the space, I will investigate the ways that the geographies that were produced through the Israeli census in turn affected subsequent depictions of the land. More specifically, the continued physical presence of subjugated populations—a presence that was acknowledged through the need to conduct a census in the first place— influenced Bachi's theoretical work in statistical cartography. Bachi claimed his findings were scientific, and that he concentrated not on political issues but rather on obtaining an objective, quantitative view of Israel and the

Palestinian Territories. Yet even Bachi's most abstruse equations were fully embedded in the cultural and material landscapes where he worked.

The Census as Western Science

Bachi had been a successful professor of statistics in Italy before the fascist government's racial purity laws forced him to flee prior to World War II. Over the course of his career, Bachi was active either in government, academia, or independent research, throughout nearly every major political transition in the region in the second half of the twentieth century, from before the founding of the state of Israel in 1948 until just before the official signing of the second Oslo Accords in 1995. From the beginning, he insisted that the Israeli census be firmly rooted in Western science. The 1948 census was carried out while the war that followed the founding of Israel still raged around them. But even then Bachi contended that the census would be conducted according to rigorous and objective statistical principles (Bachi et al. 1955; Leibler and Breslau 2005). Later, as the director of the Israeli CBS, he continuously argued that military concerns should not be allowed to dominate over scientific rigor (Bachi 1981; Leibler and Breslau 2005).

Throughout this process, he looked to Europe and North America for models for his scientific work. In the planning stages of the Israeli censuses of the 1950s and 1960s, Bachi and his colleagues requested numerous census documents from countries such as Canada, France, and Spain—and coincidentally, from Iran, which at that time was still under the control of the shah, who was politically supported by governments in Europe and North America (CBS 1969b). Bachi was a main contributor to both the *Statistical Atlas of Italy* (Bachi 1999) and the *Atlas of Israel* (SOI 1970), and spoke at innumerable international conferences (e.g., Bachi 1955, 1962a, 1962b, 1974b, 1975, 1989).

Bachi's repeated attempts to present his research as objective and neutral are evidenced by the fact that under his leadership, the CBS fought to compile data "solely according to professional considerations, without interference from political quarters" (Schmelz and Gad 1986, xii). Yet Bachi's more politicized work counting Palestinians nonetheless exerted a strong influence on his abstract research, and did so in ways that were not wholly incompatible with his aim of developing statistical cartography as a science. As a result, Bachi's work was both thoroughly scientific and thoroughly colonial at the same time. It therefore demonstrates how empirical science and settler-colonial practices can coincide.

Indeed, Bachi's academic context was also related to his governmental efforts. In addition to running the census, Bachi founded the Department of Statistics at the Hebrew University of Israel, thereby further linking academic statistics to Israeli state objectives. His two major academic books were published in tandem with census milestones: the first, in 1968, immediately after the census of the Occupied Territories discussed above, and the second, after his death in 1995, following the earliest census in Israel to fully incorporate quantitative digital mapmaking of the kind that Bachi had long advocated (Bachi 1999, ix; "In Memory of Roberto Bachi" 1996, 13–14; Schmelz and Gad 1986, xi–xiv). As I examine in detail below, both books address key issues of the census and develop methodologies first implemented in the enumeration of the OPT. As such, the theoretical, academic, and governmental areas of Bachi's professional career were thoroughly interlinked, and mutually reinforced one another.

The relationship between Bachi's governmental and theoretical work was also affected by the advent of digital cartographic technology. Throughout his career, Bachi took part in a technological transition in cartography from a professionalized trade that relied on hand-drafting skills into a quantified, mechanized science that became heavily dependent on computers. Bachi himself was trained within a modernist tradition of statistical cartography in Italy. He was instrumental in bringing the paradigm to the nascent Israeli state, where it was further developed in part due to his personal efforts (Schmelz and Gad 1986). Additionally, later methods enabled the detailed management of piecemeal territories—a type of complexity that is considered a hallmark of postmodern or late modern digital governance.

Late Modernism and the Influence of Landscape

Computers are often presented as a break or revolution in the history of technology, an abrupt disjuncture between the modern and postmodern. However, Bachi's maps serve to demonstrate that this is not an all-or-nothing transition. Rather, Bachi chose to further key modernist goals in formulating cartographic knowledge, such as accuracy and the continuity of borders, and this lent continuity to the maps he made both before and after computers became commonplace. He also actively rejected other goals considered to be hallmarks of modern statistics, like completeness and the consistency of space. So his later studies arguably are evidence of a new form of modernism that challenged the fundamental goals of completeness and consistency, but nevertheless it is one that still would be accepted and recognized among international scientific communities.

Moreover, Bachi's research illustrates how in the context of modernization processes, the landscapes that are produced can transform the goals of modernization. The efforts to enforce a single set of national borders for Israel instead spiraled into the construction of tightly linked networks of bounded settlements in the West Bank. In time, the very process of determining their borders would transform the aims and purpose of Bachi's theoretical work. Bachi largely cooperated with broader efforts to literally and figuratively expunge Palestinians from the land. Yet he could not extricate himself from the social and material connections afforded by his physical proximity to Palestinian communities. As a result, years of occupation ultimately fed back into his statistical cartography, shaping both Bachi's research trajectory and, through him, the very fabric of the digital canvas on which population statistics in the region continue to be mapped.

In what follows, I analyze key cartographic methodologies that Bachi developed, with a focus on his two major theoretical books on statistical visualization that appeared in 1968 and 1999—the latter posthumously from work completed in 1995. I show how each book was influenced, even at its most quantitative and theoretical levels, by the very presence of Palestinian populations in the landscape—an influence felt in terms of the concerns that arose while counting populations in the West Bank. It also was inflected through central political and demographic arguments of each book's respective era: in 1968, through the charged claim that Palestinians did not exist, and second, in 1995, by way of the ongoing debates that came to be known as the *demographic war*. I first introduce each book and the census that predated it (in 1967 and 1995, respectively) in the context of their related political debates. I then delineate the development of Bachi's theoretical claims in relation to the relevant census methodologies. I proceed by examining each book in the context of Bachi's work as a whole, pinpointing his simultaneous retention of modernist cartographic ideals such as accuracy and continuity, and rejection of others, like completeness and consistency (Haraway 1988; Livingstone 1992; Monmonier 1991). In the conclusion to this chapter, I return to the role of computerization in

Bachi's efforts to innovate while maintaining international scientific standards.

By analyzing Bachi's research in the theory of statistical maps in 1967 and 1995, this chapter encompasses the entire period of direct and full Israeli control of the West Bank and Gaza Strip, until the beginning of limited management on the part of the PA. This period also coincides with the early history of digital cartography. During these decades, even as Bachi sought to cement the transformation of Israeli statistical cartography into an international science, he himself was thoroughly situated in local and regional as well as international landscapes.

The Critical Cartography of Census History

National censuses are a prime means by which state bureaucrats collect information about the people under their jurisdiction. The census has played a central role in linking academic knowledge to the management of human mobility in Palestine and Israel (Zureik 2001), and cartography is crucial to the use of the census in surveillance and control (Crampton 2003).[3] Foucault's call to study geographies of power has sparked a body of research into the ways that colonial geographies were integral to modernity's core (Stoler 1995). Yet this work has only just begun to be applied to the critique of quantitative geographic knowledge (Crampton and Elden 2007). While the central role of census mapmaking is often acknowledged in the academic literature (Anderson 2006; Crampton 2003; Edney 1997), statistical cartography's technical aspects have less often been treated in detail (exceptions include Hannah 2001, 2009; Mood 1946; Pavlovskaya and Bier 2012).

Even so, the census doesn't simply take place within predetermined national borders. Instead, it forms part of the labor necessary to constitute the nation on the ground (Anderson 2006; Leibler 2004, 2007; Leibler and Breslau 2005; Zureik 2001). In studies of the census in postcolonial contexts, modern conceptions of space have taken more of a pivotal role due to the well-documented relationship between cartography and imperialism (Cosgrove 2008; Edney 1997; Godlewska and Smith 1994; Gregory 1994; Harley 1989; Kalpagam 2000; Stone 1988; Turnbull 2000; Winichakul 1994) as well as the links between the spread of technologies and bureaucratic systems for managing populations (Bektas 2000; Feldman 2008; Hull 2003;

Leibler and Breslau 2005; Mrázek 2002; Zureik, Lyon, and Abu-Laban 2011). But to date less critical attention has been paid to methods for conceptualizing census boundaries and the spatial epistemologies that are implicit in such definitions. There are notable exceptions, including the literature on gerrymandering in statistical districts more broadly (Bunge 1966; Cranor, Crawley, and Scheele 1989; Mood 1946; Sauer 1918; Sherstyuk 1998). In addition, Elia Zureik (2001, 227) has analyzed how borders have been shaped by "a Palestinian–Israeli dialectic of state construction" framed by asymmetrical power relations—although Zureik emphasizes continuities in contestation over the transformations in maps.[4]

International Hierarchies of Scientific Knowledge

Just as those in power have attempted to define and manage populations through the census, their own subjectivities have also been influenced by the geographic contexts that they operate within. Bachi certainly served as a privileged denizen of dominant statistics in Israel, but his international role was complex. He benefited from international hierarchies because unlike Palestinian statisticians, he came from a nation that was widely acknowledged as a nation. As discussed in chapter 2, Israel's dominant culture was increasingly viewed, within a global context of racism and Orientalism, as racially white and culturally European. But as a Jewish scientist from a small nation, Bachi was constrained by these same hierarchies, which favored large and established states, and also operated in a global context of anti-Semitism.[5] Thus, international asymmetries would intricately shape the ways that Bachi could innovate while still being accepted as a scientist. Such recognition was crucial to participating in international debates as well as being viewed as a theorist whose methods were credible and whose findings could be accepted as objectively true.[6] With this in mind, Bachi's insistence that the census be scientific—namely, that it be conducted not according to political dictates but rather in line with the requirements of rationality (Leibler and Breslau 2005)—was not only the outcome of personal preference or beliefs. Instead, it can also be viewed as part of an attempt to fit within an international academic hierarchy whose members often had rigid, if implicit, notions of belonging (Porter and Ross 2003; Wagner 2001).

These notions indicate one additional way that Bachi's work was constrained by international hierarchies. Bachi appears to have embraced the dominant paradigms in demographics wholeheartedly. Still, his attempts to demonstrate his scientific legitimacy are also the logical outcome of the widespread belief that there is only one true way to do science (Smith and Marx 1984; Wyatt 2008b). It has long been held that the power of science and technology is due precisely to the fact that scientific discoveries are true everywhere that they apply, independent of social or cultural influence. Yet this view masks the incredible local variation in science and its findings. Furthermore, it has important implications in terms of the international landscapes of science and technology, because it includes an assumption that the "right" way to do technoscience is that practiced in Europe and North America.

There is an implicit logic to this rather-illogical assumption, and it reads as follows: if there is one right way to do technoscience, and that right version is the one practiced in Europe and North America, then developments that do not conform to Western technoscience are unscientific *by definition.* For this reason, it would be imperative for Bachi and other Israeli academics to demonstrate their willingness to conform to dominant conceptions of science and technology in order for it to be recognized internationally that their efforts indeed were scientific. Even as he used methods that were internationally recognized to assist in controlling Palestinian populations, then, the fact that Bachi was not working solely in Europe and North America would have set limits on the scope of his potential innovations.

The Quantitative Revolution: Geography as a Statistical Science

This raises the question: what was this interdisciplinary paradigm to which Bachi strove to adhere? For starters, it was increasingly mathematical and digital. Bachi sought to computerize his cartographic methods early on, starting in the 1950s. His role as an early adapter of digital mapmaking involved countering the historic perception of the discipline. At the turn of the twentieth century, geography, including the subdiscipline of cartography, had been considered a field comparable to history. While attempts previously had been made to establish it as a science (Godlewska 1999), it was widely believed that geographers gave descriptive, analytic accounts of particular landscapes (Livingstone 1992).

This changed after World War II, as a new generation of geographers worked to transform geography into a statistical science—a movement that became known as the "quantitative revolution" (Barnes 2001). Due to their efforts, by the 1960s mapmaking was no longer viewed internationally as a descriptive spatial record of territories. Instead, as noted in chapter 2, it was seen as a scientific method for using a Cartesian grid to display variations in statistical quantities (Godlewska 1999; Livingstone 1992).[7] So by the 1960s, although Bachi himself was quite sensitive to the specifics of geographic mapmaking, he tellingly uses the words *graph* and *map* almost interchangeably. This is in keeping with the conventions of quantitative cartography, where maps are merely one type of a broader category that includes graphs of mathematical functions.

As a contributor to this international quantitative revolution, Bachi made seamless transitions back and forth between population statistics and geostatistical theory. Thus he put mathematics to work, via technology, in the name of modern empirical science. In Bachi's (1968, 1–2) case, his stress on clarity and transparency is apparent in his exhortation that GRPs would make graphing "quick and easy," and would avoid types of mapmaking that are "inaccurate or even misleading." By framing his efforts in this way, Bachi helped early on to bring the insights of geography's quantitative revolution to Israel, and serve as a link between Israel and international networks of technoscience. But since he was an innovator as well as transmitter of this new methodology, he challenged the assumption that the quantitative revolution, or indeed the concomitant development of GIS mapmaking, took place primarily in Europe and North America. For comparison, although the US census had been developing computer cartography for the purposes of census enumeration since the late 1960s, the first GIS files were only created for the 1990 census, a mere five years before Israel fully computerized its population counts.[8] As early as the mid-1950s, while the quantitative revolution was just starting up in Europe and the United States, Bachi (1956) was already pursuing similar avenues of research in Israel.

The transformations in Bachi's maps also illustrate how the quantitative revolution was not wholly a revolution. Instead, there were significant consistencies in cartographic methods before and after the advent of computers. Bachi's work highlights the specifics of this process, for although he might not have been able to challenge the paradigm wholesale, he nonetheless was able to innovate within its bounds. He navigated the alternating process of change and conformity to international dictates. Bachi adapted existing statistical methods so that they might become practically useful in the contexts, like the census, in which he worked. This process would transform Bachi as well. Through the ultimate ambivalence in his depictions of statistics for Palestinians, the Palestinian struggle for self-determination would also leave its mark on his research and cartography more broadly.

Locating Existence under Occupation, 1968

The outcome of a conflict between two parties usually doesn't hinge on a debate over whether or not one of them exists. But this is precisely the way that some Zionist groups have attempted to frame debates over the right of Palestinians to live in and enter Israel and the Occupied Territories. Prime Minister Golda Meir's famous proclamation in 1969 that, "There never was such a thing as Palestinians. ... They did not exist" ("Golda Meir Scorns Soviets" 1969), has been reiterated over countless election cycles both within Israel and abroad.[9] Meir's and others' challenges to Palestinian identity have been skillfully rebutted at length (Bishara 2003; Kanaaneh and Nusair 2010; Khalidi 2006, 2010; Massad 2001, 2003, 2006; Seikaly 1995). Yet her comments deserve further scrutiny for how they relate to contemporary debates.

In the longer piece from which the quotation is taken, Meir bases her argument on principles of statehood, claiming, "When was there an independent Palestinian people with a Palestinian state?" By suggesting that Palestinians don't exist because they have been prevented from building a nation, Meir in effect uses the language of social constructivism against itself. As suggested by the title of *Imagined Communities*, however, Anderson's (2006) foundational book on the nation, although imagination might be necessary for the formation of a nation, formal legal recognition is not. So, few contemporary researchers would join her in claiming that nationhood is what brings those people into existence.[10]

Nonetheless, Meir's assertion has long been used for instrumental reasons, and its power in this respect shows the political danger of severing the social from the material, as noted in chapter 2. Given that the social has been the focus of much recent work, in this chapter I concentrate strategically on the more material aspects of the Palestinian presence in the land and their influence. So instead of emphasizing how the social shapes the material, this chapter shows how material landscapes, which themselves are imbued with the social, also feed back into social and scientific knowledge. Indeed, through Bachi, they would come to have a decisive effect on Israeli population maps.

Palestinian Existence and the Israeli Census

Irrespective of any claims to Palestinian nonexistence, from the perspective of the Israeli military administrators who took over the Palestinian Territories in 1967, the Palestinians existed and needed to be counted. The first steps were to devise population categories, count people according to them, and note the precise locations of every individual. This is precisely what the CBS set out to do. But the population count was not only aimed at the surveillance of those in the OPT. If the push to conduct a census after the 1967 war demonstrates a tacit acknowledgment of Palestinians' existence, the census also aimed to limit the number of Palestinians who were allowed to remain. As Anat Leibler has shown (2004, 2007; Leibler and Breslau 2005), one of the primary motivations for conducting the census so quickly was to prevent those who had fled during the conflict from returning. The census was the basis for issuing identity cards that allowed their bearers to reside permanently in the Palestinian Territories, if not to become citizens. So by performing the census early, the administrators prevented those who were away from gaining the right to come back.

As a result, the census could be said to have two potentially conflicting priorities: on the one hand, to gain an accurate count of the populations now under Israeli control; on the other hand, to exclude as many people as possible in order for fewer Palestinians to be able to claim residency. Since the population of Israel was produced in part through the census, it was crucial for census takers to preemptively make as many Palestinians as possible uncountable, and therefore invisible and, for national purposes, nonexistent. For as John Law, Evelyn Ruppert, and Mike Savage (2011, 8) have pointed out, as far as any particular method is concerned, "that which is invisible for all intents and purposes doesn't exist." So it was in the best interest of the census takers both to rigorously count Palestinians who were there and shape the population by excluding Palestinians before anyone was ever counted.

In this context, the census authorities were especially wary of being charged with undercounting. They consequently stress the extreme lengths to which census enumerators went to obtain precise enumerations. Their report notes that "despite the use of special vehicles (and even donkeys), the enumerators could not reach isolated houses or distant localities (especially nomads' tents), because of difficulties of access or danger of mines" (CBS 1967, xxx–xxxii). In so doing, they highlight some of the ways that as a result of enduring political realities, much of the West Bank became practically inaccessible to the Israelis in the immediate aftermath of the war. So just as the very presence of Palestinians made the 1967 counting necessary, in the eyes of Israeli officials, the ongoing geographic impacts of 1948 and 1967 circumscribed their ability to conduct that census. The census in turn would affect the types of maps that Bachi used and advocated in his 1968 book. His theoretical academic work is thus emblematic of the census maps that had tremendous practical power over Palestinians' lives.

Bachi played a key role in the way the 1967 census was conducted, and his name appears on the official report both as the "director of the census" and "government statistician" (CBS 1967). In the process, however, he contradicted one of his own judgments from 1948. Back then, against members of the Israeli military who had wanted to count Palestinians with different methods from those used by Israelis, Bachi argued forcefully that the methodology should remain the same for both groups. He claimed that only with consistent methods would the results be seen to be statistically rigorous (Leibler and Breslau 2005). The 1948 census reports also described the hesitancy to enumerate Jews via a curfew, given that this method had been widely used by the British during their occupation after World War I, and thereby would have brought back traumatic memories (Bachi et al. 1955). Yet by 1967, under Bachi's direction, the census did precisely this for Palestinians—a process undertaken for the OPT alone. Bachi oversaw the one-day curfew with the stated reasoning that indiscriminately confining people to their homes would improve the chances of counting as many people as possible.

So the curfew was justified using arguments for accuracy, but it had its own consequences in terms of the accuracy of the statistics. The curfew indeed excluded refugees who were missing, away from home, or homeless due to the war. It also, by the census takers' own omission, had the effect of creating "differences between the locality in which [inhabitants] were registered and the permanent places of residence" (CBS 1967, xxx–xxxii). People were counted according to where they happened to be under the curfew, not necessarily where they primarily lived. The resulting census data for the Palestinian Territories would have presented a logical conundrum for Bachi. As someone who considered himself a scientist, to Bachi the Palestinian census data would have seemed less than rigorous—seeing as how they were collected under restrictive conditions that made them only partially comparable to census data for the fully annexed areas of the state of Israel. Yet also as a scientist, he could not completely ignore that the data existed— for example, by placing labels that read "no data available" on relevant areas of his maps of Israel. After all, he spearheaded the operation that collected the data in the first place. So on scientific maps, the question for Bachi would have become: how is it possible to best map Israel while neither denying that data exist for Palestinians, nor actually including that (only semirigorous) census data on the map?

The presence of the Green Line boundary between Israel and those Palestinian territories that were newly occupied compounded the difficulty of how and to what extent to acknowledge the existence of data for Palestinians. As recounted in more detail in chapter 4, the Green Line was first drawn during the cease-fire agreements of 1949, and in places between 1949 and 1967, it served as a highly militarized border between central Israel and the (Jordanian-controlled) West Bank. While Jerusalem was rigidly divided, in some areas the separation was inconsistent, and border zones, including those where homes had been demolished in the process of marking the border, were purposefully filled in with Israeli settlements. As a result, by the late 1950s, Hadawi (1957, 1) could claim that in many areas, "the Armistice Lines have not until this date been demarcated on the ground." As discussed in chapter 1, from 1967 onward even the official policy of enforcing the border was reversed. The agencies of the state of Israel claimed that the Green Line was no longer a valid international border. Instead of referring to the military occupation of the Palestinian Territories, academics and government agents euphemistically proposed that Israel and the OPT had been "reunited." In keeping with this claim of reunion, the Green Line no longer appeared on most state maps (Benvenisti 1984; Gorenberg 2012; Shehadeh 2007, 178).[11]

The official omission of the Green Line border after 1967 was in contrast to the situation at ground level, given that the Palestinian Territories occupied in 1967 have not been formally annexed by the Israeli state— with the notable exception of East Jerusalem. The administrative, military, and economic policy discrepancies between annexed Israel and the un-annexed OPT were and are extreme (Benvenisti 2000). The separate-and-unequal policies of the 1967 census is emblematic of these injustices.[12] As a result of the Palestinians' existence and ongoing struggle, for most census purposes, it was the Green Line, not the official state line, that served as a boundary between Israel and the Palestinian Territories occupied in 1967. This circumscribed the collection of data for the majority of the Israeli population.

The resulting inconsistencies in mapping the Green Line are borne out in Bachi's (1968) *Graphical Rational Patterns*. The aim of his book is to introduce the eponymous GRPs as a standardized set of symbols for displaying numbers on maps. Maps often indicate numerical values with circles of varying sizes, but the different size of the circles can be difficult to judge visually. By constructing his elaborate symbology of GRPs, Bachi hoped that anyone familiar with GRPs would be able to tell the exact numbers the cartographer sought to display (see figure 3.1). Through GRPs, though, Bachi also started to move away from rigid notions of consistency, which would have required him to display all internal borders including the Green Line, in favor of greater accuracy for a few selected points. Perhaps not surprisingly, the points he selected to display most often represented Jewish Israelis.

In fact, despite directing and overseeing the census of the OPT, Bachi rarely, if ever, used the data from that census in his work. To omit the data without appearing to be unscientific, Bachi increasingly began using one type of map from among two commonly available options. As is evident in his 1968 book, for mapping census data collected from 1967 onward, Bachi started to favor graduated circles more and more. He became increasingly critical of the shaded area technique called *choropleth*, which was more commonly used at the time. Choropleths are maps in which districts, states, or regions are progressively shaded darker or lighter in order to represent increased percentages of some particular characteristic, such as the average number of people in each household for specific areas (see figure 3.2a). Choropleths are convenient because unlike graduated circle maps, they allow cartographers to indicate statistical data without an abundance of different symbols. At the same time, they can be misleading, and there are several problems that arise. For instance, because different regions are shaded in, those with larger geographic areas often stand out as more significant than they would otherwise appear to be, if they were judged by the statistics alone—and this is only one of multiple layers of complexity.

In contrast to choropleths, graduated circle maps (see figure 3.3) are those, as described earlier, that use shapes of different sizes, generally circles, to represent a particular statistic. The difference between choropleths and graduated circles are not obvious at first, and both are widely used. But graduated circle maps have one advantage that relates to Bachi's research: they allow for the omission of certain boundaries, including the Green Line. For instead of shading an entire area or subregion, the graduated circle is located with respect to one single point on the map. This allows for greater precision in depicting—or choosing arbitrarily, since a random point could be used—the precise locations of populations within national boundaries. The result, however, is that in graduated circle maps, the actual boundaries of those districts do not have to be included on the map. In contrast, on choropleths, if each district is shaded in, then the edges of those shaded regions—their boundaries—are already implicitly indicated (figure 3.2a).

Bachi's 1968 Book: From Choropleths to Graduated Circles

The trend toward graduated circles and omitting the Green Line can be seen progressively over the course of Bachi's work. The standard practice for graduated circle maps is still to indicate boundaries, even though it's not practically necessary. Yet as Bachi moved increasingly toward graduated circles, the Green Line appears less frequently on his maps. In Bachi's (1968, 196–197, 227, plates viii and x) *Graphical Rational Patterns* book in particular, several maps of other regions or earlier data indicate the internal boundaries of the area being mapped. By contrast, nearly every single map of the region that depicts data collected in 1967 or later uses graduated symbols as opposed to choropleths (e.g., Bachi 1999). Similarly, prior to 1967, his graduated circle maps rigorously displayed the Green Line (e.g., Bachi 1962b). Then, in the first years following 1967, the boundary was irregularly omitted and displayed, seemingly without respect to whether the data displayed included those for the West Bank or not (Bachi 1974a), and over time it was dropped entirely.

Bachi's increasingly frequent omission of the Green Line was coupled with the disappearance of maps that show broader views of the region, and an increasing substitution of maps of other countries for those of Palestine and Israel. Throughout the 1970s, Bachi moved toward only mapping Israel and Palestine at a finer scale, with a focus on Jerusalem. During this period, he also expanded his use of choropleths of Italy and the United States to substitute for the absent maps of broader Israel and Palestine (e.g., Bachi 1975). But for these maps, too, he eventually came to only use graduated circles, with the result that even the internal borders of Italy and the United States would come to be alternately included and excluded from them (e.g., Bachi 1999).

In his 1968 book, Bachi still advocates the use of choropleths in combination with GRPs. He describes several methods of doing so, thereby showing that choropleths were not technically or rationally incompatible with his GRP (Bachi 1968, 198–216). Bachi nevertheless criticizes choropleths in the same work. He claims that when using choropleths to represent percentages of population by subregions or provinces, "the distortion resulting from this method may be extremely dangerous" (ibid., 205). Later, comparing a choropleth and graduated circle map with GRPs for the same areas, he notes that "the visual impression obtained by the two graphs [maps] is completely different and almost opposite." The graduated circle map "enables us to receive the correct impression" while the choropleth "may thus fail almost completely to convey an *accurate* view of the distribution under survey; Moreover, comparison of data for each region and of the national average may create an impression of discrepancy." He argues that this discrepancy is not present in graduated circle maps that use GRPs (ibid., 210–213, emphasis added). The practical result of these convictions was that Bachi's maps were both increasingly accurate, by pinpointing values via graduated circles and GRP symbols, yet increasingly inconsistent in terms of his depiction of the Green Line.

In his 1999 book, the Green Line is absent from every one of Bachi's relevant maps. This includes both his multiplicitous maps of the Jerusalem metropolitan area (which straddles the Green Line) as well as his sparse maps of Palestine and Israel more broadly (Bachi 1999, 46, 104, 149, plates 2 and 3). But even when the Green Line is not indicated, a ghost of the political border can be seen on maps where areas of high Palestinian concentration, depicted as blank areas, in effect sketch out the boundary as a type of palimpsest (see figure 3.3). Choropleths were not entirely absent from Israeli census publications after 1967. Although the Green Line was

Figures 3.2a and 3.2b
At left is a sample choropleth map from the 1961 Israeli census. At right is a map that shows the extent of the development of an Israeli national GIS database as of 1996. In figure 3.2a (CBS 1963), the shaded areas indicate the average number of people in each household in that subdistrict, called a "natural region," as defined by the CBS. Because the map was made prior to 1967, data for the West Bank and Gaza Strip are not included. Reprinted with permission. In figure 3.2b, the shaded regions indicate those areas where digital map data existed or was in progress. The SOI appears to have prioritized the West Bank and Gaza Strip, which both fall under the "Mapped" category—namely, those areas that were converted first to GIS from paper maps.

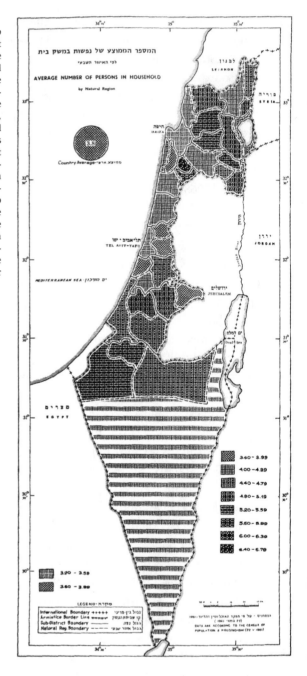

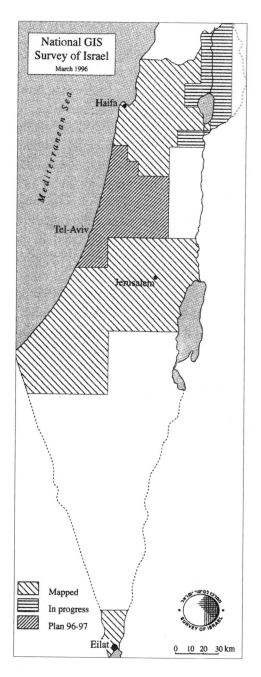

National GIS Survey of Israel
March 1996

Mediterranean Sea

Haifa

Tel-Aviv

Jerusalem

Eilat

Mapped
In progress
Plan 96-97

0   10   20   30 km

Figures 3.2a and 3.2b (continued)
Notice how the use of shading requires the cartographer to display some internal boundaries—at the very least, between the areas that are different shades. Here the Green Line is the boundary of the West Bank and Gaza Strip, indicated in black. Yet even if the Green Line were not indicated, the boundaries of the Israeli districts shown fall along it. So simply shading those districts gives an impression of the Green Line. Figure 3.2b demonstrates how, in part but not solely due to Bachi's legacy, GIS data in Israel were developed using an overhead view that seamlessly incorporated all of the Palestinian Territories (Peled 1996). This was likely accomplished precisely with the expectation that Israel would lose the ability to collect data on the ground throughout the entire Palestinian Territories due to the handover of limited sovereignty to the Palestinians following the Oslo Accords. It contrasts, but also coincides, with the shift to data that included only Israeli areas within the Palestinian Territories rather than the Palestinian Territories as a whole (figure 3.4).

Israel. Total population by 1055 settlements and cities, 1992: (a) map based on circles: ·1011 • 134,700 • 249,800 ● 356,900 ⬤ 556,500; (b) map based on GRP: . 1011 ▪ 10,118 ✔ 19,224 ⁛ 50,590 ■ 556,500 + (Source of data: List of Locations, Israel Central Bureau of Statistics, 1993).

Figure 3.3
A later map of Bachi's, comparing graduated circles (left) to GRPs (right). The Green Line and any indication of the PA are omitted. Yet even though Palestinian cities and towns are not shown, a blurry outline of the West Bank can still be seen. This is because the omitted Palestinian towns are concentrated in the West Bank, given that many Palestinians are not permitted to live in central Israel (see Bachi 1999, 165). Reprinted with the permission of Springer and Kluwer Academic / Plenum Publishers, *New Methods of Geostatistical Analysis and Graphical Presentation: Distributions of Populations over Territories*, chap. 6, "Graphical Rational Patterns," 165, Roberto Bachi © 1999.

often left off its population maps, the CBS did continue to use choropleths sparingly on technical maps that depicted census boundaries. These were intended primarily for internal use, and were aimed at Israeli and/or Jewish statisticians (Bachi 1974a; CBS 1969b, 1985). Yet unlike his earlier work (Bachi et al. 1955), by the late 1970s Bachi was no longer routinely using CBS choropleths, even though technically such maps were still available.

Over time, as Bachi came to avoid indicating the Green Line on his statistical maps, this also limited his options in terms of data visualization and display. As discussed above, he justified these limitations on scientific grounds, while perhaps showing some uneasiness. The omissions, however, conveniently fit the practice among state Israeli cartographers of no longer depicting that border. So instead of having two common options available to him for mapmaking, both choropleth and graduated circle maps, Bachi was restricted to using only graduated circle maps. This restriction had significant implications because it dictated which statistics might be visualized. Moreover, it affected the entire course of Bachi's subsequent research. It is perhaps then not surprising that his GRP symbology consists precisely of an innovation almost exclusively in the area of graduated circle cartography.

If by creating GRPs, Bachi was attempting to construct a more refined system of statistical visualization, then it was obtained at the price of the consistency in the representation of regional territorial borders. In addition to addressing contemporary concerns with the visualization of statistical data, by grappling with the Green Line, Bachi's work is representative of how the existence of the Palestinians and the occupation more broadly shaped the key methodological issues that Israeli cartographers faced in the unique post-1967 context. But Bachi's efforts with graduated circles and points didn't require him to give up mapping borders altogether. Instead, by 1999, amid Israeli settlement construction in the context of the ongoing demographic war, Bachi would begin using points to map entirely new boundaries of his own.

The War of Scattered Boundaries, 1999

If the 1967 census represented tacit acknowledgment of the existence of significant populations in the newly occupied territories, the 1995 census showed the persistence of the demographic war—a competition between groups, whose members are encouraged to bear ever more children in order to secure greater political power (Courbage 1999; Kanaaneh 2002). It is the obstetric expression of the adage "strength in numbers." On the Israeli side, its proponents peddle fears that Palestinians outnumber Israelis. They compile evidence that European-origin Israelis are a minority in Palestine and Israel, which frightens those concerned with maintaining Israel's status as a Europeanized nation. They thereby highlight the currents of racism and discourses of racial purity in the conflict (Massad 2003).

Bachi himself was a central actor in demographic debates and is credited as being one of the earliest scholars to warn of (allegedly) impending Jewish population decline around the world ("In Memory of Roberto Bachi" 1996; Schmelz and Gad 1986). A consideration of Bachi's later work, in light of the 1995 census's attempts to quantify the outcome of almost thirty years of demographic struggles, provides a fitting counterpart to the analysis of the claim of Palestinian nonexistence that was explored above. From one perspective, the demographic war is the flip side of the assertion that Palestinians did not exist. While the existence claims mistakenly suggest that there were no groups present that (legitimately) identified as Palestinian, in contrast, the rhetoric of the demographic war argues that Palestinian numbers are so high that they threaten to "engulf" (non-Palestinian) Israelis. The current prime minister, Benjamin Netanyahu, has even referred to a "demographic bomb" (Munayyer 2012). Seen from another perspective, however, the demographic war and existence claims have much in common. For in statistical terms, the question of whether a group *exists* is often translated to a question of whether that group has a high enough population to be considered statistically *significant*. And the census is precisely the mechanism to determine which groups have numerically significant populations. So both the question of whether a group exists, along with the question of which group has or will have the largest population are determined by the census.

Israeli Settlements in the 1995 Census

As with the existence claims, then, what is striking about the demographic debates are the interlinkages between governance methods, social discourse, and the material bodies of a population. For although Meir's statement that Palestinians didn't exist, as noted earlier, was framed in historical terms it could also be considered an exercise in prophesy. It is only possible to claim that Palestinians didn't exist in a situation where many Palestinians had been violently expunged from the landscape. To turn it around, Meir's claim about Palestinians nonexistence in the past was more a prediction that she would not let them build a nation of Palestine in the future. Despite this, on Bachi's later graduated circle maps, the erased Palestinian populations are so numerous that they show up precisely through their absence, as a white, blank area. For example, in figure 3.3, the Green Line and any indication of the PA are omitted. Even though Palestinian cities and towns are not shown, a blurry outline of the West Bank can still be seen. This is because the omitted Palestinian towns are concentrated in the West Bank, given that many Palestinians are not permitted to live in central Israel.

So after World War II, there were organized and concerted efforts to manufacture the "fact on the ground" that Palestinians weren't a numerically significant population by preventing those who were pushed out from returning (Leibler and Breslau 2005; Morris 2004), and attempting to ensure that they would not continue to exist in years to come. Yet Meir's claim was bolstered by further action. Zionist leaders worked to bring Jewish families to Israel, in part to counter any potential claims that Jews themselves did not comprise a large population in the region. They encouraged emigration from around the world. They also transported large numbers of Jewish families to Israel from elsewhere in the Middle East in the 1950s, in the context of the broader regional protests over the founding of Israel, and eastern Europe in the 1990s, following the fall of the USSR (Meir-Glitzenstein 2011).

After 1967, this was coupled with coordinated attempts to bolster the Jewish population in the Palestinian Territories through the settlement movement, which was inspired by fears that the significant Palestinian population of the OPT—including large numbers that had been made refugees in 1948 and had subsequently resettled in the territories—would result in Israel losing its hold on the region. As a response, Israeli settlers began moving to segregated enclaves in East Jerusalem and the Palestinian Territories. The aim was to complicate any future geographic division as well as obfuscate what until 1967 had been a starkly separated territory (at least with respect to central Israel) by building a small and diffuse but tightly linked network of Israeli settlements (Weizman 2007; see chapter 5).

As a reaction to fears of Palestinian expansion, the settlements therefore also represent a tacit acknowledgment of Palestinians' existence. Nonetheless, although the settlements were created as an effort to make the material world conform to claims made about it—and specifically to confound traditional methods of drawing boundaries between groups—the resulting changes to the population landscapes would influence maps of Israel. In the context of the ongoing drive toward the West Bank, the CBS and Bachi began to develop methods that would allow them to draw borders around complex, small population clusters such as the settlements. Likewise, Bachi's innovations in his later work lay precisely in his ability to separate what might otherwise have been inseparable: to draw out and count the intentionally imbricated Israeli settlers from their distributed points across the West Bank.

Yet it was not a foregone conclusion that the settlement populations would be counted in the Israeli census, given that they were located on disputed territory where, as we have seen, Palestinians were separately enumerated. The need for the CBS to give the state legitimacy by demonstrating the numerical strength of Israelis, however, dovetailed (perhaps unsurprisingly) with state efforts to claim the settlements as being an integral part of Israel. In contrast, the publicly available Israeli counts of Palestinian populations were left deliberately vague. After 1967, the CBS did not attempt to publicly count Palestinians in the OPT again—that is, except for those residing in areas annexed to Israel, such as East Jerusalem. The CBS instead relied on estimates of population growth (e.g., Ministry of Defense 1973, 98). Moreover, after the 1994 signing of the Oslo Accords, Palestinians gained some limited sovereignty, and the duty of counting Palestinians was transferred to the newly formed PA. For this reason alone, the 1995 Israeli census was unique. It was the first census after 1967 that was conducted with the expectation that the Palestinians would also conduct their own census, focusing on Palestinian-controlled areas of the OPT as well as East Jerusalem at roughly the same time. The Palestinian census was in fact completed in 1997 (PCBS 2008).[13]

In the context of the demographic war and new Palestinian census, the 1995 census marked a push forward for increased accuracy on the part of the CBS methodology. Although the CBS conducted censuses roughly every decade after 1967, the 1995 census was a methodological watershed of sorts. For starters, it was the first census to comprehensively use GIS mapmaking in an attempt to record the precise location of "every" dwelling—and thereby, in theory, every person in Israel (Lasman 1997). In this respect, it built on the groundwork of the SOI, which had started converting its paper maps to a digital GIS framework in the late 1980s (Peled 1996). Moreover, instead of incorporating digital maps on the side, the census shifted wholesale toward conducting the census *through* the use of GIS. Districts were redrawn, and the counting process itself was mechanized, as GIS was used to determine and print maps for each enumerator to follow while conducting the count. In addition, a national geographic database was developed to store census data for the foreseeable future (Bahat 1997; Barak 1997; Ben-Moshe 1997a, 1997b, 1997c; Blum 1997; Calvo 1997; Kagan 1997; Lasman 1997; Peled 1996; Stier 1997).

Distinct parallels can be drawn from the methodology of the 1995 census and Bachi's later work. Bachi himself did not have a direct hand in the run-up to the 1995 census because he had retired from the CBS, although he remained professionally active until he passed away in 1995 ("In Memory of Roberto Bachi" 1996). But in many ways the 1995 census represents the outcome of the methodology that Bachi developed over the course of his career. Bachi was influential in all the major developments that culminated in the 1995 census. These included the early turn to computerization, emphasis on meeting international scientific standards, and efforts to rely on direct counts versus smaller statistical samples to pinpoint exact locations of individuals and form a "snapshot" of the population (Leibler 2004, 2007).[14] Indeed, much of the material for Bachi's 1999 book was written by Bachi himself precisely during the years that the 1995 census was in development.

Bachi's 1999 Book: Convex Hulls from Point Features

In the context of the demographic war, the CBS wanted to count as many Israeli Jews as possible. So the question would have become not whether but instead *how* to include the settlements in the census. In the end, an attempt was made to seamlessly incorporate the post-1967 settlements into the new computerized regions. In the 1995 census, each settlement is defined as one or more census tracks, and they appear throughout the CBS's publications, including lists of "towns in Israel" and maps of the "urban areas of Israel" (CBS 2000b), without reference to their unique status as settlements within the Palestinian Territories (figure 3.4). Similarly, the

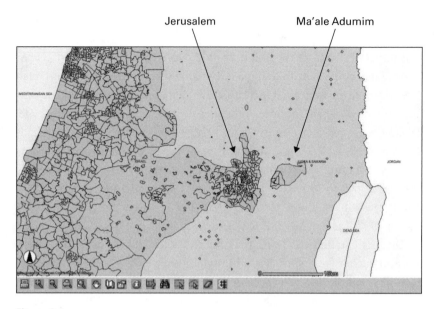

Figure 3.4

An annotated screenshot of Jerusalem (center) from the CBS online interactive map of 1995 Israeli census data. This map was made in the context of the Oslo Accords, and the Green Line is drawn clearly and prominently through the center of the map. It surrounds the (Israeli-defined) Jerusalem metropolitan area and divides the annexed areas of Israel on the left from the West Bank on the right. Yet the West Bank is referred to using the adapted biblical name of "Judea and Samaria"—a use that is part of a political strategy to avoid reference to the Palestinian West Bank. The map shows the borders of statistical areas, revealing how Israeli settlements in the West Bank (the smaller pockmarked areas at right) are incorporated as Israeli municipalities for the purposes of the census. The largest of these areas, just to the east (right) of Jerusalem, is the settlement of Ma'ale Adumim. The attempt to incorporate Israeli West Bank settlements into the census has meant giving up on the contiguity of Israeli territory. Here, Israeli territory includes all the complex areas toward the left as well as the settlements toward the right, which are depicted as numerous islands of Israeli sovereignty within the West Bank. The settlements appear in far fewer numbers than their present levels, however, because the map, made in the mid-1990s, does not show more recent expansion. Annotations by the author (CBS 2000a).

properties of Israeli settlements in the West Bank are registered with the Israel Land Authority rather than Israeli administrators of the Palestinian Territories. This helps to legitimize their status, legally and economically, as "Israeli" lands (Shehadeh 2007, 83). In Hebrew, the word *settlement* (*yishuv*) is most frequently used in its general sense to refer to all places of steady human habitation. So throughout the CBS and other Israeli state maps at the time, no internal borders are shown, no Palestinian towns in the OPT are shown, and there is no distinction between Israeli cities in Israel (which are also referred to as *settlements* in the more general sense) and the Israeli settlements in the Palestinian Territories.

The 1995 census therefore indicates an ongoing shift in the conceptions of sovereignty at work, from one that seamlessly incorporated all of the West Bank (as suggested by figure 3.2b) to one that omitted Palestinian-controlled areas but nonetheless selectively retained the Israeli settlements (figure 3.4).[15] Whereas Palestinian populations were treated entirely separately by the census in 1967, using distinct definitions and methods of enumeration, in 1995 the groups of Israelis in the same land are fully incorporated, despite the fact that they lie beyond the Green Line. Furthermore, while the Green Line was not often displayed explicitly, the census was still actively involved in drawing implicit and explicit boundary lines. For the settlement boundaries were being redeveloped and reinstated with increasing sophistication. The definition of these settlement boundaries was a lengthy and detailed process (Calvo 1997) that required new methods of the sort that Bachi just happened to be developing at the time.

In his 1999 book, the culmination of his life's work, Bachi avoids choropleths altogether. But this does not mean that he didn't emphasize boundaries. Yet rather than focusing on political boundaries, Bachi centers his arguments on methods for drawing complex statistical borders around existing small population groups. These are precisely the types of methods that would have perfectly fit the needs of the Israeli census because they would allow the census to use statistics to contend that the settlements fit within the "statistical area" defined by the Israeli population. However, this effaced the process whereby—instead of producing statistics in order to count existing population distributions—the settlements were put in place precisely so that they could be counted.

One of Bachi's methods is particularly useful in this context: his research on convex hulls, a type of mathematical set. The problem the census faced was how to define the national area in light of population groups, like the settlements, that were not geographically contiguous with the main area of Israel. Convex hulls represent one possible solution. The *convex hull* is a mathematical term for a boundary that fits tightly around all the points of a specific set. If there are a finite number of points in that set, then forming the convex hull is like stretching a rubber band around the outermost points (figure 3.5).

Finding a convex hull would be especially useful, for example, in determining the areas that are defined by close networks of Palestinian towns—and thereby determining the leftover packets of space that might be settled by Israelis. It also parallels the redrawing of political boundaries after 1967. Jewish settlements were founded in key strategic spots, and then political boundaries were redrawn around those points in a sort of convex hull. This was justified by noting that since Jews now lived there, the areas were to be incorporated into the Israeli state. Most notably, such borders include the expanded boundary of "Greater Jerusalem" that accompanied the Israeli annexation of East Jerusalem and large areas of the West Bank that immediately surround the city. Analyzing redistricting in relation to the demographic war, Meron Benvenisti (1999) has claimed that "the annexation boundaries did not determine the city's demographic ratio. Rather, the 'optimal demographic ratio' has created the city's boundaries" (see also Zureik 2001).

Yet in time, the material efforts of those redrawn borders would shape the social conceptions that had inspired the borders in the first place. For example, instead of using one solid political boundary, whether for nations or cities, the use of convex hulls provides a more textured and complex rendering of scattered populations. This rendering was made possible in part through the settlement project. For the convex hull is also useful for estimating the geographic size of individual settlements, then statistically appending their areas to the total area of Israel. Indeed, although settlements have political boundaries, the inhabited areas often expand quickly—a process that involves claiming outside lands for defense walls and other security structures (Weizman 2007). Thus the populated area of any particular settlement, which is of interest to a census concerned with mapping every habitable building, not infrequently falls outside the municipal area, and attempts to deal with this situation would mold settlement cartography.

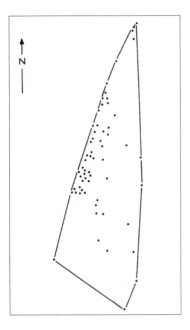

Figure 3.5

A map of Tel Aviv hotels (points) with an enclosing convex hull (line). This figure demonstrates how point or circle maps can obviate the need to show political boundaries. None are given, and the image has been made at a high level of abstraction. In addition, it illustrates how Bachi begins with data and then constructs statistical boundaries—in this case, a convex hull consisting of the straight line that connects the most distant points.

The Palestinian city of Jaffa, which falls within the broader Tel Aviv–Jaffa municipal boundary, is cut off from the map. This is accomplished through the act of drawing the convex hull, thereby transferring the background area of the hotels from the Tel Aviv municipal (political) boundary to the convex hull (statistical) one. The resulting boundary is entirely dependent on the accuracy of the statistical data set that is used, which appears to have only included Israeli-owned hotels. Bachi (1981, 1018; 1999, 91, 94) defines the convex hull as the "gross [territorial] range," yet here it serves precisely to exclude Palestinian residents. For an analysis of how Tel Aviv was defined and built through a contrast with—as well as the obliteration of—Jaffa, see Rotbard 2015. Figure reprinted with permission.

Bachi's 1968 book is evidence of how, in his early work, he started omitting internal boundaries and moving from areas to points. In contrast, by 1995, Bachi had largely abandoned his attempts to rectify GRPs with choropleth maps. He turned instead to drawing boundaries around statistical data, beginning to constitute new borders based only on those points that he selected as being significant to the census. In so doing, Bachi again adapted and innovated on international statistics in order to meet the strategic political goals of the Israeli state. In the process, he rejected the traditional focus on the consistency of national territory in order to retain the continuity of borders around the fragmented territories, including the settlements, claimed by the state. His methods allowed for an enlargement of the total area of Israel that is inhabited by Israelis, and they did so in a way that naturalizes the settlements as a part of Israel.

In addition, as noted above, when the total area of a particular settlement is determined based on the furthest points of habitation, these can be far larger than its official municipal boundaries, which might take longer to catch up to the pace of construction. Any private Palestinian areas that happened to fall between two outlying points would simply be incorporated into Israeli territory through the use of the convex hull—that is, by drawing an enclosed line around those (newly) Israeli points and calling everything in between "Israeli." In combination with GIS software, the geographic database that was developed in the course of the 1995 census enabled the cartographers to calculate convex hulls for hundreds of thousands of individual points at ever finer scales, including individual buildings, thereby defining and quantifying a set of ever more multiplicitous borders. The convex hull method sacrificed national contiguity in order to maintain strict and increasingly numerous boundaries. Yet the fact that there were more borders did not mean that they became more open or less guarded; quite the contrary.

Conclusion: The Continuity of Computer Cartography

Population statistics, like those developed in the national census, form one of the core connections between the state and local communities, and they are central to everyday governance practices. As one of the primary means through which population statistics are communicated, census cartography fundamentally influences both national and international public imaginations of those governance practices as well as society itself. Yet cartographic methods are also shaped in local social and geographical landscapes. Bachi's attempts to make graphing and statistical mapmaking more readily available served to place him in the international statistical vanguard, but his work continued to be informed by the political realities of the Israeli occupation.

This was not obviated by the spread of computers and their allegedly universal rationality. In the end, instead of representing a radical departure from previous cartographic methods (November, Camacho-Hübner, and Latour 2010), computers and GIS proved useful in legitimizing Bachi's political position. While the public use of his GRP symbols was fast outpaced by computer graphics capabilities—even to the extent that they appear less frequently in Bachi's (1999) own later research—his contributions to core concerns of geographic and demographic statistics, and their application in statistical cartography, continue to be influential (Louder, Bisson, and La Rochelle 1974; Shea and McMaster 1989; Wegman and Carr 1993).

Contrary to popular rhetoric at the time, the advent of computers did not unilaterally involve further abstraction from local landscapes. Indeed, computers represented the triumph of rather than a challenge to the theoretical foundations of Bachi's work, including his efforts in local and regional landscapes. They also allowed for the incorporation of the physical existence of Palestinians and the Israeli settlements, which in turn influenced the main directions of population cartography. In 1967, the very presence of Palestinians on the newly occupied areas meant that the census cartographers, including Bachi, had to rush to count and manage them. The policy of omitting the Green Line, in the context of the existence of large numbers of Palestinians in the landscape, presented a challenge to Bachi's attempt to construct a cartographic image of the homogeneous nation of Israel. As a result, he began phasing out his use of shaded choropleths instead of exclusively using graduated circles in combination with his GRP symbols. In 1999, the ongoing conflict along with the settler movement that was a response to the continued Palestinian presence meant that the census now had to include disparate and noncontiguous settlements in tabulations of Jewish populations. Consequently, Bachi would further innovate on his use of dots by adapting convex hulls for geostatistical use. He thereby moved from developing ways of mapping that obviated the need for borders and toward finding ways to draw continuous statistical boundaries around noncontiguous political spaces.

By omitting internal boundaries altogether, Bachi could include Israeli data without making plain that such data for the Palestinian Territories were either not being collected or, in the case of the census, not being indicated on the map. He then helped to actively naturalize the presence of Israeli settlements in the landscape by redrawing boundaries around the area of Jewish habitation—an area that by design, were spread across most of the Palestinian Territories. As such, both GRPs and convex hulls were useful in constructing facts that were empirical and political. Yet in order to reconstruct a total national area that includes the settlements, the 1995 census first had to extricate the settlements from their surroundings— that is, precisely the numerous Palestinian towns of the West Bank and Gaza Strip.

Bachi's research is therefore both fully internationally scientific and thoroughly influenced by local political contexts. Despite his best efforts to position himself otherwise, Bachi's (1980) work even at the most abstract levels was shaped by the landscape of Palestine and Israel. At the same time, his choices were circumscribed by the hierarchies of international technoscience—whereby his legitimacy depended on his ability to present his findings as objective and exact. Computers lent credence to this goal. But the tension between maintaining supposed empirical rigor, on the one hand, and adapting his methodology to the people and landscapes so that they might further the practical goals of the Israeli government, on the other hand, is one that continued through the adoption of GIS in the region. These tensions played out in different ways in the nascent PA, which is the subject of the next chapter.

4 The Colonizer in the Computer: Stasis and International Control in PA Maps

The Steadfast Colonial Legacy

Political struggles in Palestine and Israel have long been thoroughly international.[1] The founding of Israel and war that followed took place in concert with the end of British colonialism in 1948. Yet British rule was legitimated not only through military conquest but also by a mandate from the League of Nations, the precursor of the United Nations. Over the years, the US government has played a central role in negotiations to end the conflict as well—and been criticized for failing in its attempts to be an impartial arbiter, in no small part due to ongoing military aid to Israel. however, military funding was not the only form of aid. Though smaller by comparison, waves of humanitarian assistance have flowed—from the United Nations and EU member states, among others—into the PA, which was established through the Oslo Accords of the mid-1990s. With them have come a host of scientific and technical experts, offering training to local professionals in the hope that international science and rationality could help to solve ongoing disputes. While military aid is viewed as an attempt to deepen the conflict, to support one group at the expense of another, humanitarian assistance is believed to be part of an effort to help bring into being a recognized, rational, and technocratic Palestinian state.[2]

Despite aspirations that international assistance might provide a corrective for military occupation, the confluence of British and Israeli imperialism has shaped the PA's efforts to establish a distinctly Palestinian form of empirical knowledge in the West Bank. In this chapter, I show how occupation and colonialism have reconfigured the practice of scientific and technological expertise. Drawing on the concept of steadfastness, or *sumud*, I argue that the material legacies of colonialism can and do work through forms of internationalism.[3] In the process, they circumscribe cartographers' attempts to develop *stasis*, which I define as the ability to remain in place.

The study of stasis is crucial within two broader contexts. First there are the debates over the unjust effects of the Israeli occupation, which too often focus solely on immobilities due to checkpoints, walls, and road-blocks, for example, without also investigating how immobility relates to the enforced mobilities of the occupation. Second, there is the study of global capitalism more broadly. Though uniquely related to Palestine and Israel, enforced mobilities are an integral part of global capitalism, and this requires a more thorough engagement with the relationships between stasis and mobility. Maps are one emblematic case of how dominant international knowledge, and technoscientific knowledge in particular, is grounded in regimes of both steadfastness and movement.

In the late 1990s, PA cartographers developed a highly sophisticated system for making maps using GIS cartographic software with the goal of developing a national mapping program, thereby helping to build the nascent Palestinian state in keeping with nationalist ideals (Tesli 2008). However, their work took place in the context of a continued international presence in the region over the course of centuries. As such, it interacted in complex ways with the legacy of British imperialism as well as the ongoing Israeli occupation of the West Bank and Gaza Strip. Yet although the connections between maps and nationalism are well known in Pales-tine and Israel (Leuenberger 2012), there is a further necessity to analyze cartographic methods in addition to the completed maps. In what fol-lows, I focus on these methods and their associated practices. Studying the implications of stasis for knowledge practices, the chapter's aim is to broaden and deepen both mobility studies and critical research on tech-nology in Palestine and Israel. I first look at the stasis of maps from the British Mandate period in order to demonstrate how the PA's extensive use of British maps as data sources in turn affected the scale and resolution of early PA cartography. The chapter continues with an analysis of maps made in the period 1994–2000, from the founding of the PA to the start of the second widespread Palestinian uprising, known as the Second Inti-fada, against the Israeli occupation. In particular, I show how the use of British maps led to broad depictions of the West Bank that were less useful for daily governance. I then turn to the techniques that PA cartographers and institutions have used to build stasis in the period 2000–present, in the context of the Israeli influence on PA mapmaking through the ongoing occupation and the maps it helps engender. Thus this chapter deals with the relations among three stages of mapmaking, from the British Mandate period (1920–1948), to the early implementation of GIS (1994–2000), followed by the proliferation of digital maps in daily governance (2000–present).

To complement its state-building motives, the PA also produces knowledge as a way of countering the Israeli occupation (Romani 2008; Zureik 2001, 227). Although the 2006 election of Hamas increasingly drew the Israeli military toward the Gaza Strip, at present the occupation continues to severely limit the scope and extent of places where the PA can collect data in the West Bank. In the process, it leads to the continued erasure of rural areas from the map. By concentrating on PA cartographers in relation to the omitted people and areas, I point to the difficulties of even privileged attempts to expand the purview of dominant forms of knowledge. This provides an interesting complement to the observation of Sally Wyatt, Andrea Scharnhorst, Anne Beaulieu, and Paul Wouters (2013, 2) that there is a "cyclical feedback loop" between new forms of knowledge and new infrastructure. For in the West Bank, the restrictions—and at times direct attacks—on infrastructure have hampered the development of new and innovative ways of producing knowledge.

British Mandate maps have been hailed as an empirical triumph (Gavish 2005), but even so, the British colonial authorities emphasized hard boundaries in their efforts to define territories across several scales—from attempts to control private ownership by marking out land parcels, to proposals to divide the region into two separate states. This has been criticized as contributing to the beginnings of conflict in the region (Fischbach 2011; Weizman 2007, 14–15). The legacy of hard boundaries is evident even in the text of the 1994 peace treaty that allows for the establishment of "permanent, secure, and recognized" borders between Israel and Jordan "without prejudice to the status of any territories that came under Israeli military government control in 1967"—that is, after the start of the Israeli occupation of the West Bank and Gaza Strip. This focus on stability is especially striking in relation to the allowances that were built into the same treaty to account for the fact that this particular border follows the ever-mutating course of the Jordan River (Hashemite Kingdom of Jordan and State of Israel 1998). There is a certain irony, then, that the ongoing colonial legacy has led the PA to strive to end the occupation by making use of the very colonial maps that may have helped to start it in the first place.

Building Stasis and Sumud

Yet if PA cartography must be viewed in light of British and Zionist colonialism, then it also takes place in the context of efforts to build sumud among Palestinians in the region. The concept and practice of sumud, which Lori Allen (2008, 456) refers to as a "nationalistically inflected form of stoicism," incorporate notions of rootedness and perseverance. Sumud has come to play a key role in antioccupation struggles, not least because it offers a "third way" between hatred of and acquiescence to oppressors (Shehadeh 1982, viii).

First coined at the 1978 Arab League Summit in Baghdad, and intended primarily as a means of raising funds abroad to support Palestinian communities, in time sumud began to be used to draw attention to the frustration felt by many in the absence of political gains. This in turn had the effect of highlighting the work that is necessary for Palestinians to remain in place on the land, both pragmatically and metaphorically. Following the writings of Raja Shehadeh (1982), sumud currently refers to a spectrum of activities that promote steadfastness and constancy in the Palestinian Territories and beyond. Contemporary examples of sumud range widely, from the ongoing act of emphasizing the social and cultural contributions of Palestinians in the face of efforts to deport, assassinate, or otherwise remove them, to active support for the health, education, and related infrastructure that makes group identity as well as physical survival possible (van Teeffelen and Giacaman 2008). It is also referenced by organizations such as the Sumud Story House in Bethlehem and the Sumud Choir.[4]

There is a broad literature on the political theory and everyday practice of sumud such as, to cite a few examples, works by Nadia Abu-Zahra and Adah Kay (2012), Tahrir Hamdi (2011), Rami Isaac (2011), Craig Larkin (2014), Mohammad Marie (2015), and Leonardo Schiocchet (2011). Some have been critical of sumud's role in shaping resistance in the Occupied Territories because the practice of sumud favors a form of self-organization—in the sense that all are expected to work in alignment with the common goal of supporting Palestinian communities—that does not fully conform to dominant notions of a formalized political agenda. For instance, Jeff Halper (2006), the cofounder of the influential Israeli Committee against Housing Demolition, has written that sumud is a "non-strategy" because it is not programmatic and predetermined. By way of contrast to the expectation to provide an overarching strategy, however, Lena Mhammad Meari (2011, 2014) analyzes sumud in prison interrogations with reference to Deleuze's notion of "singular revolutionary becoming." Rather than an attempt to fix a plan or group identity in advance, this suggests that sumud's very success may come from its tendency toward openness and multiplicity.

Building on sumud, here I explore one very specific aspect of it: the sense of staying put, of working to sustain a presence in a particular landscape. In order to distinguish my concept from sumud's more multifarious connotations, I call it *stasis*, as noted above. Stasis is not equivalent to the absence of movement, but rather it can result from the tension between equal yet opposing forces. Although it might possibly have negative connotations of stagnation or being stuck, stasis also brings with it a positive sense that deserves greater attention than it has had to date: namely, that of generative stability achieved in relation to ongoing change. I aim to reclaim these positive connotations while incorporating the sense of geographic context that is implicit in the term. Thus, as it is used here, stasis refers to all the effort necessary to establish and maintain a presence within a spatial context in addition to the effort necessary to maintain the related landscape.

Stasis is in no way unique to the maps of PA cartographers, yet its expression in the context of the occupation has a marked currency that makes its production explicit in ways that might otherwise lie dormant. My focus on stasis, however, is in no way meant as a corrective for or means to supersede sumud. Indeed, because it has a long history in Palestinian movements, I make no claim to rethink sumud as a broader practice, leaving it to those working within those movements. Instead, out of respect for practices of sumud, I pull out one specific aspect that they bring to the fore: stasis. This is precisely in order to avoid claiming ownership over the use of sumud while, simultaneously, thinking through stasis in relation to academic literatures that are explicitly concerned with the relationship between international knowledge regimes, space, and landscape.

Even so, the study of stasis does complement research on Palestinian relationships with borders and land. These include the analysis of the cartographic practices of "figuring it out" that Linda Quiquivix (2013) examines in her work on Palestinian cartography. In addition, scholars of mobility have considered stillness and immobilities (Cresswell 1999, 2011; Kotef 2015; Söderström et al. 2013), and an illuminating study has been done of Palestinian discourses of mobility and virtual mobility (Aouragh 2011a). Yet stasis is not simply equivalent to either stillness or immobility. Often those who are not granted the ability to move are simultaneously denied the opportunity to remain—a logically impossible position that affects countless refugees. In different locales, the relationships between stasis and mobility can vary widely, then, at times forming a dialectic, and at others, overlapping in diverse, but specific ways. These relationships therefore are deserving of further ethnographic research across contexts.

Mimi Sheller and John Urry (2006, 210; Söderström et al. 2013, 10) have noted that mobilities are predicated on "often highly embedded and immobile infrastructure" and view them as "multiple fixities or moorings." This conceptualization can be further enriched by the long tradition of urban geographical analyses of how infrastructures are produced and institutionalized. To give an example, stasis might refer to a small business owner's ability to maintain a particular food stand—including the work of, say, paying the rent or mortgage, obtaining permits, updating the interior and exterior, ordering and inventorying goods, and perhaps defending the area from violent attacks. Yet it could equally refer to efforts to preserve a neighborhood or region in the face of indiscriminate developers, industrial pollution, or military conflict. In this vein, Bas Van Heur (2010, 125) has emphasized the interaction of obduracy and processual change, allowing that the city is always "becoming but also being, movement as well as stasis, circulation as well as sedimentation."[5] A study of stasis also demonstrates how attention to the city and land can serve to further broaden mobility studies to include the effects of geographic and political landscapes on both stasis and mobility.

While mobility is usually presumed to occur on top of spaces that are presented as a mere background to movement, PA cartographers don't simply travel through preexisting territories.[6] Rather, they move through spaces that themselves change and are changed by their movement—a form of material feedback that contrasts with philosopher Ian Hacking's notion that material objects do not react to the categories attributed to them. Thus, stasis does not require reference to the movements of particularized individuals or a Cartesian grid. Instead, it involves ongoing work in changing material and social fields. As Allen (2008, 156) observes in reference to sumud, in "conditions where the routine and assumptions of daily life are physically disrupted ... everyday life in Palestine—in its everydayness—is itself partly the result of concerted, collective production." Noting the connection between the home and conceptions of homeland among Palestinians in Jerusalem, Amahl Bishara (2003, 144) draws on the writings of Arjun Appadurai to indicate the challenges of dominant categorizations that "leave some people trapped, and others placeless altogether." So rather than indicating an absence of movement or change, stasis draws attention to the work of maintenance and care that is required to remain in place in social and material landscapes—including everything from political negotiations to the upkeep of buildings—all while helping to construct that place as being the *same*, as existing continually over time.

Both mobility and fixity can lead to disempowerment, but through a focus on the production of place, stasis provides an additional aspect for the study of the knowledge created within political and social movements. Stasis in emerging states is especially pressing in an age when all nations are expected to be *international*. Nations are required to "play nice" with other nations by participating in supranational institutions like the United Nations, International Monetary Fund, and World Bank; establishing formal diplomatic relations; and engaging in negotiations. Yet as work in decolonial and critical development studies has shown (Escobar 2011), such participation is Janus-faced. On the one hand, further integration into international hierarchies and systems of power can enable access to material support and international aid. On the other hand, it can make the state subject to the sometimes-debilitating terms and requirements that are often conditions of that aid. So instead of assuming that diplomacy is positive in its own right, greater focus is required to understand the effects of specific forms of internationalism on the stasis of postcolonial communities.

The Shortest Distance between Ramallah and Oslo

Discussions about the role of sumud were reinvigorated after the 1993 Oslo Accords provided an international legal mandate to found the PA, giving it a limited jurisdiction in the West Bank and Gaza Strip. Although they led directly to the establishment of an administrative body for the Palestinians, and as such represented a concrete step toward self-government, the Oslo Accords came to be viewed as a partial betrayal of the struggle for sumud, both in terms of the concessions granted to the Israelis and due to the jockeying among different PLO factions at the expense of concrete gains (Parker 1999; Said 2001).[7]

Framed as interim agreements, the first Oslo Accord in 1993, and second agreement in 1995, were geared to set the stage for full Palestinian sovereignty in the Occupied Territories. The ultimate boundaries and possible land swaps were intended to be determined by a final status agreement— one that has yet to materialize despite several rounds of negotiations, including the failed Camp David and Taba talks of 2000 and 2001, respectively. However, after the election of the Islamist party Hamas in 2006, the members of the more pro-Western and economically liberal Fatah party, which was in charge of the PA, refused to step down. This succession crisis resulted in a split between Fatah and Hamas, evidenced most starkly in the PA's closure of the parliamentary body, the Palestinian Legislative Council. As a result, the PA continues to rule by decree in the West Bank (UNOCHA 2010a), while Hamas maintains a parallel government in the Gaza Strip.[8] Nonetheless, a new rapprochement was achieved in 2017, and political rifts have been overshadowed at times by the subsequent wars and ongoing blockade that continues to prevent basic foodstuffs and materials from entering Gaza.

Despite the varied critiques of their impact, the Oslo Accords have also come to be accepted as marking the founding of PA cartography. This highlights both the internationalism of the PA as well as the fact that official Palestinian mapmaking was equally geared toward a specifically international goal: to influence the ongoing political negotiations. In adopting this frame, the PA drew on a long history of international cartography in the region. The legacy of Ottoman and Arab regimes have often been downplayed or even actively erased in Palestine (Tamari 2009), yet empirical mapmaking efforts that involved Palestinians go back at least one hundred years. They include the Ottoman efforts of the nineteenth and twentieth centuries (Foster 2013; Gavish and Ben-Porath 2003), Egyptian maps of the 1950s (Gavish 1996), and Jordanian program of land registration (Jordanian Department of Lands and Surveys 1966).

These international efforts in the Arab world were increasingly met with related maps produced within Palestinian nationalist movements. Maps were hand drawn in the service of the Palestinian cause at least as far back as the 1940s. The PLO in the 1980s (Gavish 1996) made similar efforts—and these generally omitted reference to Israel just as Israeli maps frequently failed to show the Palestinian Territories, as discussed in chapter 3. They also include numerous attempts to document the locations of demolished Palestinian villages and cultural geography (Abu Sitta 2004, 2007; Hadawi 1970; Khalidi 1992). But it was not until the 1990s that the Negotiations Support Unit of the PLO began developing its own maps to counter the Israeli ones then in joint use in the negotiations— maps that not surprisingly, had been uniquely suited to defending the official Israeli position (interview 6, former Physical Planning and Institution Building [PPIB] project cartographer). The Oslo negotiations thus are routinely cited as the first time that the Palestinian representatives brought their own maps to the table (interview 6, former PPIB cartographer; Quiquivix 2013). Palestinian contributions to cartography multiplied in the ensuing years, and extended into digital topographic modeling (Weizman 2011, 68).

After Oslo, many of the Negotiations Support Unit cartographers would go on to work for the PA in various mapmaking capacities (Tesli 2008), but their role in negotiations does not mean that empirical efforts were not addressed in their early maps. Quite to the contrary, as the Israelis had long demonstrated, the way to gain political clout internationally was to produce maps that attempted to represent objective truths in a form that would be internationally recognized. Therefore, the cartographers demonstrated their legitimacy to international monitors by advancing empirical evidence for the nature and location of particular claims to land (Sayegh 2000). As a result, from the beginning, Negotiations Support Unit cartography was both avowedly international and thoroughly national. It also aimed to be fully digital (Sayegh 2000; Tesli 2008; interview 6, former PPIB cartographer)—an ambitious goal that was in keeping with the simultaneous trend to computerization in the Israeli state.

The focus on digital cartography was in part due to the well-known difficulties that arise from the use of paper maps to determine boundaries (interview 6, former PPIB cartographer). One example is the often-retold story of the birth of the Green Line boundary of the West Bank during the 1949 negotiations to end the war that broke out in 1948. The border was drawn by hand onto a paper map, and on that paper map it seemed thin and detailed. However, when administrators later tried to mark the now-militarized boundary on the ground in the Jerusalem region, they found that large areas of the landscape lay entirely inside the line of the boundary, whose thickness at its most refined was equivalent to 250 meters, or the width of several buildings (figure 4.1) (Cameron 2011; Blake 1995;

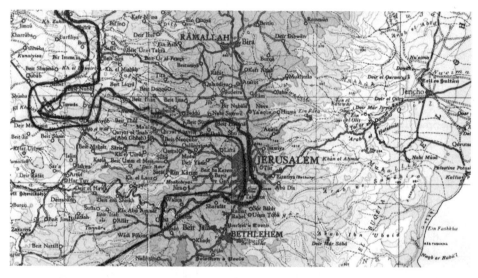

Figure 4.1
An excerpt from a 1:250,000-scale British map of Jerusalem with hand-drawn border overlay, taken from the original map where the Green Line boundary of the West Bank was drawn in wax pencil during cease-fire negotiations in 1949. This map well illustrates the perils of paper maps. Here the "Green" Line is the right-hand line that forms part of the double line that begins near the top-left corner and then heads vertically through Jerusalem before leaving the frame at the bottom left. The line varies considerably, crossing roads and villages at will, and covers the eastern side of Jerusalem almost completely (SOP 1949). It is a double line because it also demarcated a "no-man's-land" between the opposing forces. This map is the responsibility of the author. For updated UN maps of Palestine and Israel, see http://www.ochaopt .org/mapstopic.aspx?id=20&page=1 (accessed April 2, 2016).

Brawer 1990).[9] This created a challenging situation for villages that literally fell within the borders.[10] During interviews (interview 6, former PPIB cartographer; interview 1, Palestinian cartographer and NGO director), this story was used to indicate the inherent error, as it was claimed, of paper maps. If viewers zoom in on a paper map, or a fixed image of one, the details quickly become blurry. In contrast, digital maps like Google Maps redraw in greater detail with ever-finer zooming—although there is still an upper limit.

The inconsistencies found in paper maps did not disappear with the advent of digital cartography. They instead were transformed in ways that while arguably more accurate, did not necessarily lead to greater power or control. After the Oslo Accords, Palestinian cartography continued apace and the use of GIS began to spread through the higher echelons of the PA. This is evidenced by the early success of the Palestinian Geographic Information Center, the first major PA GIS agency (Abdullah 2005). After the center's brief tenure, it gave way to an additional effort that was essential to the process of bringing GIS to the Palestinian state. The PPIB project was funded by the Norwegian Agency for Development Cooperation with the aim of building long-term cartographic infrastructure in the West Bank, including servers, modems, and digital storage (Tesli 2008). The PPIB project is central to the history of the PA's efforts in Palestine because it was later integrated with urban planning and the Ministry of Planning (MOP), and thereby became one of the main mapmaking initiatives whose work is currently in use in daily governance in the West Bank. Historically it is significant given that it lasted from 1995 to 2008 (Tesli 2008). This includes the period immediately following the Oslo Accords, on through the Second Intifada, and then the split between Fatah and Hamas. The PPIB thus provides a central site to analyze stasis and mobility throughout the formative early period of the PA.

The PPIB also affords a unique opportunity to study the role of international agencies in the PA's daily functioning. For rather than simply representing a failure of Western democracy and a related push away from the West, the PA's maneuvers to oust the democratically elected Hamas have brought it closer to the European and North American powers. Thus PA cartography was made possible by an international accord, and then the backers of that accord sustained it as the PA overthrew the laws that brought it into existence. This makes it all the more urgent to investigate the ways that international landscapes influenced and helped to maintain specific forms of development and spatial knowledge.

Additionally, international political and economic development in Palestine is sometimes taken to work in opposition to the Israeli occupation. Yet this is not necessarily the case. Development and occupation can readily go hand in hand. For example, in the context of twentieth-century South Africa, Peris Sean Jones (2002) demonstrates that apartheid advocates supported, rather than opposed, calls to (separately) economically develop segregated Bantustans. These apparently progressive efforts were instead a way of legitimizing the young states as token, subservient powers that worked in concert with—and not opposition to—the apartheid regime. In this light, any critique of either the PA or the Israeli occupation must include a closer examination of the role of variegated international knowledge forms in helping to shape the trajectory of authoritarian rule.

An analysis of the PPIB also contributes to broader debates in the history and sociology of science. The efforts to produce stasis in the PA can be seen as an inversion of the historical process described by Steven Shapin and Simon Schaffer (1985, 79), whereby "the naturalization of experimental knowledge depended upon the institutionalization of experimental conventions." In the PA, by contrast, naturalized forms of knowledge have supported efforts to establish particular conventions and institutions, and those institutions then become legitimized by referencing that knowledge. The power of international interests should not be overstated, however. For even as academic theory can be used to analyze social practices, PA cartographic practices have the power to shift academic conceptions of technoscience. Among other things, PA efforts demonstrate the vast potential for heterogeneity even within dominant conceptions of science and technology.

Moving Mountains of Data: Mobilities Studies under Occupation

In the early stages of the project, PPIB leaders struggled to implement their goals at a time when restrictions on the movement of Palestinians in the West Bank were tightening and becoming ever more complex. As noted in chapter 3, in the decades between the 1967 war and 1994 Oslo Accords, Israeli forces focused on erasing the vestiges of the Green Line border from the land (Brawer 1990). In the post-Oslo period, by contrast, the focus turned toward regulating the movement of Palestinians between the fragmented areas that were given over to Palestinian management (figure 4.2).

After Oslo, although the PA was granted limited sovereignty, Israel retained control over the majority of the West Bank. The Israeli-controlled territory has become known as Area C and surrounds the numerous smaller Areas A and B, which respectively were placed under full and partial Palestinian control. This led to a mushrooming of Israeli checkpoints throughout the West Bank with the aim of monitoring Palestinian travel between the different patches of Areas A and B. Since Areas A and B are islands within the broader Area C, transit between these islands inevitably involves entering Area C as well. Trips that once had taken twenty minutes instead stretched to hours, requiring Palestinians to pass through dozens of checkpoints, where they were routinely detained, often indefinitely and without charge (Amnesty International 2003; B'Tselem 2004; UNOCHA 2009, 2012b). In addition, as the Second Intifada accelerated, and attacks and counterattacks began anew, the Israeli forces used Palestinian suicide bombings of Israeli civilians as a justification for raids on civilian Palestinian populations in combination with closures of the Occupied Territories as one entire unit. Workers from East Africa and Southeast Asia were increasingly brought to Israel to substitute for the labor of Palestinians from the OPT, who previously had formed a central subset of workers in Israel. Overall, Palestinians from the Palestinian Territories were increasingly prevented from entering the bulk of Israel (Stamatopoulou-Robbins 2011, 60).

This context of mobility restrictions contributed to a recognition of the significant role of mobility in social justice struggles—an awareness that predates the recent turn in social theory toward the study of mobility. Nonetheless, mobility studies has proven quite useful to researchers who highlight the discriminatory and often-debilitating effects of the amorphous, and itself also mobile, network of Israeli checkpoints on the rhythms of everyday life (Hanafi 2009; Zureik, Lyon, and Abu-Laban 2011). For example, a recent report outlines the severe impact that the near-total closure of Gaza has had on academia in the Gaza Strip (Gisha 2012; Nagra 2013). At times, the Israeli authorities withhold funds from the PA in reaction to political developments, such as the PA's 2012 bid to be

Figure 4.2

A map of the West Bank that indicates Areas A and B, which are under full or partial Palestinian control, combined in light gray. It also indicates Israeli-controlled Area C and natural reserves, combined in dark gray. The Red Sea is shaded the darkest gray, at the bottom right. This detail is excerpted from UNOCHA 2011b. This map is the responsibility of the author. For updated UN maps of Palestine and Israel, see http://www.ochaopt.org/mapstopic.aspx?id=20&page=1 (accessed April 2, 2016).

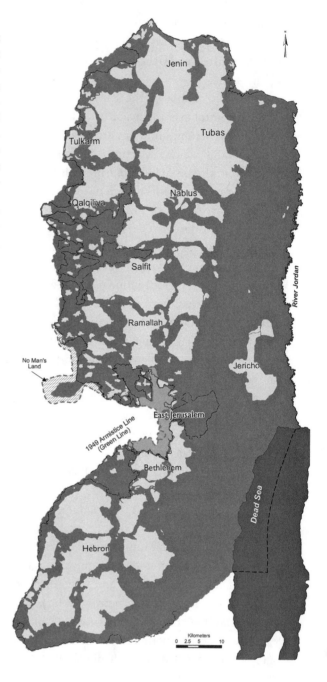

recognized by the United Nations, and this can lead to closures of public services and government offices, including West Bank universities (Guarnieri 2013). Mobility studies scholars have questioned the assumption that predefined nations or societies should be the natural unit of analysis for social inquiry (Cresswell 2011; Hannam, Sheller, and Urry 2006). They also have drawn attention to the production of breaks and stops in movement (Aouragh 2011a, 2011b; Cresswell 1999; Söderström et al. 2013; Urry 2002) and challenged the common notion that a high degree of mobility is always desirable (Vannini 2011).

Furthermore, mobilities researchers have opened up new methodologies, such as mobile ethnographies (Urry 2002), focusing on moving places along with the cultures of circulation of people and goods (Lee and LiPuma 2002). They therefore critique static conceptions of social theory. To some extent their work complements critiques of professional cartography, which is arguably far better suited toward mapping static political territories than mobile populations (Crampton 2001, 2010; Crampton and Krygier 2006). In addition, mobility studies parallels the STS and Internet studies literature, which increasingly has stressed the movement, adaptation, and differing uses of technologies across space (Aouragh 2011a, 2011b; Tawil-Souri and Aouragh 2014; Todd 1995), partly in an effort to bridge the more specific case studies that to date have characterized the discipline of STS (Sismondo 2010). Researchers have also criticized the conception that the Internet naturally grants greater discursive mobility to those who use it. Thus, as Helga Tawil-Souri and Miriyam Aouragh (2014) note, "Palestinian 'internet spaces' are grounded in offline materialities" such that they can perpetuate, rather than alleviate, colonial subjugation. This chapter complements their analysis of how the offline shapes the online, because it contributes an understanding of how offline materialities are themselves understood through digital methods like cartography.

The account of the PPIB project offered here serves to highlight two ways that the study of stasis contributes to the mobilities and related literatures. First, if mobility is to be taken as the strategic foundation for critical inquiry, then it must be accompanied by investigations into practices of stasis. As mentioned above, stasis is not simply immobility or the opposite of mobility. The ability to stay in place—and the work necessary to produce the ongoing presence of people in particular places—is not always identical to the work it takes to be accorded the ability to move. So mobility and stasis do not necessarily form a dialectic. Likewise, stasis is not identical to stability or durability, given that forms of mobility can also be stable or unstable, durable or mutable. Thus studies of mobility are incomplete if stasis is merely taken as mobility's opposite and not studied in depth in its own right.

Second, the analysis of stasis must involve an exploration of how places as well as people and objects are produced as static in the first place. Stasis does not necessarily refer to a single point or area that is timeless and unending. It instead is one aspect of the production of space—specifically, of the entire complex of imaginations and practices that are necessary to produce and maintain particular places as enduring *places* (see chapter 2). This includes the production of geographic areas or scales (Smith 1992) as enduring entities, including the national scale (Balibar 1990), and development, in practices of governmentality, of a stark separation between conceptions of landscape, on the one hand, and the people who inhabit them, on the other. So efforts to remain "in place" go hand in hand with conceptions of space, and this requires further investigations of the relations between the two.

PA Stasis: Anything but Fixed in Place and Time

The stasis of many Palestinians could be seen as something of an inversion of colonial thinking that sought to fix indigenous populations in space and time, in part by stifling international migration (Clifford 2001, 2003, 57; Fabian 2002). Instead, under the occupation, international mobility is enforced as a form of exclusion. This is sharply at odds with the more positive connotations of some discussions of migration and mobilities. In contrast to other groups, many Palestinians moved into diaspora believing they were but temporary refugees, and yet have so far been denied the option of returning. So rather than being afforded a right of movement to counter a past of colonial entrapment in place, since the early twentieth century, Palestinians have been forced to move abroad, and been denied the ability to return or remain.[11]

In this context of enforced mobility, the PA's attempts to build Palestinian cartographic infrastructure required it to reshape landscapes of science and international aid that have actively omitted Palestinians in the past (Khalidi 2010; Kreimer 2007; Massad 2006). Similar efforts occasionally have been characterized as a misguided attempt to reproduce outmoded forms of nationalism due to a type of "false consciousness," in Marxist terms, or the unreflexive internalization of colonial norms (Fanon 2008). While this critique is certainly relevant at times, such a response treats modernity as a monolithic entity and denies its complexity. Despite an ongoing awareness of the paradoxes of their position, such that they're asked to use modern knowledge in order to "solve" problems created by modern knowledge, PA cartographers do not always have the luxury of openly questioning the scientific disciplines in which they are active. For their very legitimacy depends on their ability to demonstrate an understanding of those disciplines' norms. Nonetheless, this hardly means that their participation in international knowledge arenas is somehow given or predetermined. Cartographers instead routinely find ways to modify and implicitly question prevailing paradigms. They do so in part through their very insistence on their own stasis as professionals participating in international scientific practices and debates.

The analysis that follows focuses on the efforts to produce maps that would support daily governance efforts.[12] Prior to 2008, these were concentrated in the PPIB, and from 2008 on, they were transferred to the MOP and related NGOs. Throughout, I look to provide nuance to debates over scientific practices in development contexts. West Bank elites are sometimes depicted in a bifurcated fashion—either as brainwashed adopters of Western values or, alternately, heroes of resistance to Israeli rule.

As opposed to either demonizing or valorizing local elites, I follow Saba Mahmood (2005) by attempting to move beyond conceiving of agency in terms of this dialectic between accommodation and resistance. Even as the Israeli occupation remains the dominant influence over daily life, the PA must contend with multiple forms of modernity, including international science and aid regimes. Likewise, although resistance to the occupation is a major driving factor in PA politics and maps, internal pressures and practical necessities that exceed the simple binary between accommodation and resistance have also affected the Authority's practices. In many cases, one program might convincingly be interpreted as evidence of both accommodation and resistance—or neither. Similarly, different practices or types of maps may have different implications at different geographic scales, even as they help to produce particular scales, like that of the West Bank, as more salient than others.

To better understand the relationships between international science and political sovereignty, it is necessary to study both the force and complexity of contributions to technological knowledge and practice. So while the PA has been rightly criticized for silencing dissent (e.g., Hass 2012), here I focus on the intricacies of the challenges that are faced even by elites who have the power to steer major debates.[13] In the process, I also contribute an analysis of the ways that landscapes condition knowledge practices beyond a dialectical frame. Hence I attempt to take seriously the PA cartographers' use of contemporary technologies, without presuming either wholesale acceptance of liberal scientific ideals or a purely instrumental manipulation of them for political ends (Swedenburg 1995). Doing so draws attention to how PA cartographers enact an awareness of the politics of particular epistemes even while they shape and are shaped by them (e.g., Sayegh 2000). In addition, it highlights the theoretical and material innovations they must make in order to build stasis—a process that itself requires them to rework professional landscapes that have, often quite literally, excluded them.

All the Data That Remains: The Stasis of British Mandate Maps

"There are people who would have burned them." This is how Aziz Shihadeh described his efforts at the Israeli National Library to preserve and catalog the books of Palestinians that the Israelis had seized after 1948 (Amit 2008, 20n19). Speaking to Israeli researcher Gish Amit, Shihadeh goes on to relate how some thirty thousand books were taken and diligently added to the newly founded National Library's collections (Amit 2008; Bekker 2013; Mermelstein 2011). Reading his words, I was struck by the ways that the combined theft and preservation of the lost books highlights the selective nature of what was destroyed in 1948, and what was allowed to remain. Scores of Palestinian villages were demolished in what were exceptionally tumultuous times. Some were intentionally dynamited (Khalidi 1988; Morris 2004; Pappé 1992), some were allowed to go to ruin (figure 4.3), and still others were inhabited by Jewish residents such that original tile work can be seen while walking the streets of West Jerusalem. Throughout the landscape, village ruins poke out of the dirt almost everywhere you look, from terrace walls, to pointed arches, to piles of pockmarked rubble overgrown with weeds (Benvenisti 2000). In this context, the laborious preservation of

Figure 4.3
A picture of ruins in the former Palestinian town of Lifta, near Jerusalem. Photo by
the author.

thousands of Palestinians' books is at the very least more charitable than
some of the alternatives. Yet it pales in comparison with another act of
conservation: the detailed preservation of colonial maps.

The British gave up their formal control in the region in 1948, and the
very same building that housed the SOP was rechristened the SOI. The SOI
continues to be the seat of Israeli state cartography, illustrating some of
the continuities between British and Israeli rule (Gavish 2006; Gavish and
Adler 1999).[14] Mandate rule was, as ever, obsessed with maps, and resistance
to British rule extended to the theft of maps. This was evident as the British
forces prepared to pull out of the region. Before leaving, they vainly
attempted to divide the existing maps, field notebooks, and related documents equally among the Jewish and Arab representatives. They hoped to
give Arab leaders in the city of Ramla the maps of areas that were slated to
fall under their control. They planned to do the same for the Jewish leaders
in Tel Aviv.

Yet one British report from 1948 recalled the failure of the efforts to deliver the maps in this manner: "As soon as the removal began the office was twice raided by Jewish extremists who took away a large quantity of printed maps, office machines, instruments, etc. The Army was therefore called in to remove immediately the Arab records under escort. Subsequently there were further thefts." The report also notes that "Arab raiders" stole documents and maps from the Ramla station—perhaps they didn't trust that they were slated to fall into their own hands—as well as from offices in Haifa and Nazareth (Mitchell 1948). Ramla, including the maps intended for the Arab administrators, was eventually taken over by the Jewish forces and today lies in central Israel.[15]

In the end, the majority of the maps ended up in Israeli hands. In large map collections like that of Tel Aviv University, British maps and documents spill over onto the floor. In every pocket of space above, below, and beside the library shelves, there are overflowing boxes of colonial regulation books with handwritten annotations, file cabinets full of detailed village surveys, and portfolio folders of British field maps with purple Arabic overlay, to give just a few examples. The sources include the index of a list of regulations for the railroad from 1924, apparently in use in Mandate Palestine. It lays out guidelines for people traveling with everything from barrel organs and bath chairs to automatic lung-testing machines. It also details rules for transporting large quantities of ham—with no little irony in a region where many follow Halal and Kosher restrictions on the consumption of pork (Government of the United Kingdom 1924). Everywhere I stepped during my archival research, it seemed I had to avoid tripping over British documents.

So the villages were destroyed but the maps remain. This outcome was by no means assured, given that both Palestinian and Jewish forces violently opposed the British prior to 1948. Even after 1948, reams of British papers were sold in bulk for scrap (Amit 2008), and as late as the 1980s, some documents may have been incinerated due to a lack of space—a common concern in archival collections the world over (interview 13, Israeli geography professor). Yet in the end, thousands and thousands of maps were kept and cataloged, and they have turned out to exhibit greater stasis than the very territory they sought to map.[16] In contrast, many Palestinian villages were destroyed or left unpreserved, partly because those who had the knowledge to maintain them were not allowed to inhabit them (Benvenisti 2000; SOP 1945, 4).

In addition to the fact that colonial maps have greater stasis than Palestinian knowledge forms, they exhibit greater mobility. Mandate maps of historic Palestine, including the bulk of what are now Palestine and Israel, are widely available both through online retailers and antique dealers internationally. The academics, cartographers, and PA officials that I interviewed repeatedly emphasized that the easiest part of the cartographic process was buying original or detailed copies of Mandate maps. As one PA official put it, in terms of information, "If you have the means, if you can pay, then by the end of the day you can get it" (interview 6, former PPIB cartographer). The SOI itself sells copies of British maps online, and has offered a number of historical maps through mail order for many years.

Numerous libraries, among them Israeli ones, strive to maintain public access to SOI maps, at least for those who are allowed to visit Israel. One Palestinian professor who had worked with the PPIB project remarked that in the early years, the cartographers who could enter Israel would occasionally consult the maps housed in the public collections there, while others would travel to libraries in Jordan and Istanbul in order to obtain and digitize British maps (interview 5, Palestinian geography professor). Additionally, copyright law made it more practical to use Mandate maps. In contrast to recent SOI maps, many of which remain confidential, the British maps are not only declassified but also effectively in the public domain, given that they were originally copyrighted by the Mandate government (SOP 1945, 4), which no longer exists.[17] While the colonial maps are available, the amount of work needed to acquire maps should not be overlooked. Nonetheless, they can be acquired for those with money and other financial privileges.

In recent years, the preservation and circulation of British maps thus have come to have a concerted effect on attempts to make maps for the nascent Palestinian state. During the PPIB project, the PA drew time and again on British sources, in part simply because they were there. To some extent, then, the value of the maps' scientific and technological knowledge is assured through a self-fulfilling prophecy. Knowledge forms are preserved because they are assumed to be valuable, and then they become valuable precisely because they were preserved. It is therefore essential to better understand how colonial legacies, which are materialized through such acts of preservation, continue to shape the production of knowledge.[18]

A Digitized Mandate

To begin to see the impact of British Mandate maps, it is necessary to examine the specific digital methods that were used to incorporate them into recent cartography. Given the digital focus of the PPIB, one of the first tasks the cartographers undertook was to scan the British maps into digital image files. But scanning was not sufficient to produce digital maps. After scanning, the PA employees needed to *digitize* the image files. To digitize an image, a cartographer simply uses the mouse to trace the relevant features of the scanned image, allowing the ArcGIS program to record the length, position, and direction of every small curve or area. A cartographer might open an image of a road map, for example, and then arduously draw over those roads one by one, thereby incorporating its information into the computer. Even for rudimentary maps, digitization thus is a time-consuming and grueling process. One cartographer noted that "we started from zero. … We had two employees spending eight months working 24 hours a day to digitize the contour lines of the British 1936 maps" (interview 6, former PPIB cartographer).

So the PA's high-tech mapping industry began with professionally trained cartographers laboriously tracing colonial maps. While it is not unusual for cartographers to draw on existing maps using digitization, the extent of the use of British maps was far greater due to the restrictions imposed by the Israeli occupation. This contrasts with the development of GIS in Israel in the mid-1980s and early 1990s, when data from existing maps were routinely verified by field surveys using handheld GPS—a privilege afforded to Israeli cartographers even within the Palestinian Territories due to the occupation (Peled 1996, 492). Yet as well as being time consuming, digitization selectively influenced the digital files that were produced—not least because the user decides which parts of each map to include and which to omit. Furthermore, it is not possible to digitize layers that are not present on the original map. The type of map that was used, then, also conditioned the data that resulted.

In the case of the PPIB, colonial maps shaped the digitized data in part through the specific set of geographic scales that were originally used by the British to make their maps. Geographic scales are integral to the resolution, or amount of detail, that is contained in any map. A finer-scale map shows a higher level of detail over a small area, such as a neighborhood, while a broader-scale map would show little detail but could depict a much larger area, such as an entire continent. Yet scale is far from incidental. Map data are collected at a particular resolution and detail level, and as such, every data file has a geographic scale built into it. For a paper map, the scale at which the map was originally made represents the highest level of detail of the data it contains, so that boundaries and features become chunky and fuzzy when zoomed in much closer than that range. As noted earlier, in digital maps it is possible to change between scales— literally zooming in and out, from one scale to another. However, computers can only make the transition appear seamless. To achieve the zooming effect, it is necessary to cobble together a patchwork of different maps and data sets that are often collected at different scales, different times, and under different conditions. So because they were tracing data made at fixed scales, the PA maps were restricted to the same scales that those maps used.

Of course, digitization did not preclude the PPIB from using other types of historical maps as sources. Indeed, as the project progressed, the PPIB would have used Israeli maps in particular. Even so, these also drew on British maps as data sources, especially in the West Bank, where much of the Israeli data were classified such that they were left out of public Israeli maps. So the use of Israeli maps actually helped to cement the British influence. Other types of maps that could have been used were US, French, German, Russian, and Ottoman maps, to name a few, as well as aerial photographs. Nevertheless, in addition to generally being more expensive, more likely to be under copyright, and harder to find, these maps alternately focused only on Jerusalem, or depicted the entire territory of historic Palestine at the broadest scale possible. They thus were made at even more restricted sets of scales than the British maps. Even many of the extant aerial photographs— at least those that were declassified by the late 1990s—were in fact taken by British military recognizance missions during World War I, conducted in conjunction with the British cartographic surveys, and themselves would have served as sources for the British maps (Gavish 1996, 2005, 2006; Srebro, Adler, and Gavish 2009).[19]

Therefore, whether using British or British-influenced maps, through digitization alone the PA cartographers could not improve on the scale of the data. They couldn't make more detailed features appear if they weren't there in the original map—such as in figure 4.4, which displays little to no detail in the shaded area for Jerusalem. Instead, they could simply try to trace the data as accurately as possible. The British controlled a territory much larger than the PA does, so only the finest of their common scales was detailed enough to show local features in the West Bank. The PA's extensive use of British data meant that it would be quite difficult for PA cartographers to make maps at any scale that was finer than the entire West Bank. In fact, early PA maps replicate precisely the finest British scale and no finer (MOPIC 1996, 1998). As a result, they were useful for symbolic political purposes but were less helpful for daily use. The persisting dominance of

Figure 4.4
A broad-scale British map. Zooming in on the map might enlarge the shaded area at the center, but it cannot increase the level of detail (SOP 1947). Finer-scaled Mandate maps did exist for Jerusalem and many coastal areas (see figures 4.5a and 4.5b), yet the British never completed such detailed maps for the area that would become the West Bank. Taken from the Micha Granit Map Library, Faculty of Geography, Tel Aviv University, and used with permission.

particular scales, even in digital form, contrasts sharply with the irregular stasis for PA institutions, as I explore in the following section.

Staying Put in Palestine: PA Stasis

At 11:30 p.m. on December 5, 2001, the office of Dr. Hassan Abu-Libdeh (2001a) released a statement: "The Israeli military forces have just occupied the premises of the Palestinian Central Bureau of Statistics (PCBS). Scores of heavily armed soldiers have entered the building by force." The raid lasted for six hours. Local security personnel and others in the building were detained and interrogated (Abu-Libdeh 2001d; Claudet 2001). Such disruptions appeared to be secondary to the main goal of the raid, which was to gather data. Abu-Libdeh, then the head of the PCBS, the PA department that produces the Palestinian census, told a reporter that "What I fear most is the magnitude of the data that were stolen" (Claudet 2001, n.p.). He worried that Israel would continue raiding offices to gain access to the geographic and population data the PA collects, while at the same time destroying their means of storing those data and turning them into maps (Abu-Libdeh 2001a, 2001b, 2001c; Claudet 2001). By preventing them from fulfilling their mandate, the raid could be seen as an attempt to disrupt the stasis of the PCBS not only during the raid itself but also in the future.[20]

The Israeli seizure of Palestinian knowledge centers was not a new phenomenon at the time. Since 1967, Palestinian knowledge repositories have repeatedly been subject to challenges to their institutional stasis. The Palestine Research Center, affiliated with the PLO, was subject to forced closure in the 1980s, and its records were impounded (Jiryis and Qallab 1985; Orient House 2001b). Elia Zureik (2001, 212) notes that after the Israeli invasion of Lebanon in 1982, the papers of the PLO were transported wholesale to Israel, although some of them were later returned, presumably after the Israeli military had made copies. Yet in 2001, the raids began in earnest in the context of the attacks on civilians during the Second Intifada. In May 2001, the Israeli military had taken over the Orient House, which had long served as the de facto PLO headquarters, and in that capacity, had been producing maps without GIS since the 1980s (interview 1, Palestinian cartographer and NGO director; Orient House 2005; Tufakji, n.d.). The staff members claimed that their entire archival collection was impounded, as were dozens of computers (Orient House 2001a, 2001b, 2001c, 2001d, 2001e). The twenty-five thousand volumes of the Palestine Research Center were returned two years later, whereupon the center moved to Cyprus (Jiryis and Qallab 1985). At the time of this writing, the Orient House remains closed.

The PPIB was also affected, although it managed to preserve much of its geographic data, and thus further the stasis of its project, by decentralizing its collections and moving them to undisclosed locations. As with the PCBS and Orient House, the MOP, which would come to house the products of the PPIB project, was raided by the military (interview 6, former PPIB cartographer). In this context, former participants and associates of the PPIB project routinely mentioned fears that their amassing of data might be in vain (interview 1, Palestinian cartographer and NGO director; interview 6, former PPIB cartographer). However, a PPIB employee explained that they had altered their method of data storage after the PCBS raid, and the military found little to no data in the planning offices. "When the attacks [on PA ministries] started, we took all of the data, all of the data, and we disappeared" (interview 6, former PPIB cartographer). This encouraged decentralization, requiring individual PA offices to manage their own servers— a decision that by increasing the power of both local and regional PA branches, had the potential to reconfigure the stasis of the PA central offices.

Immobile Elites: NGOs on the Move

Since 2008, the raids on institutions (e.g., Matar 2012; Nusseibeh 2012) and attacks on data infrastructure (Tawil-Souri 2011b) have continued. The incursion into mapping and census offices thus also forms part of the destruction of Palestinian knowledge and forms of information far more broadly (Aouragh 2015). Presently the stasis and physical integrity of Palestinian knowledge centers are far from assured. This adds texture to existing debates over the relevance of local scientific practices to broader institutional systems (e.g., Lynch 1993, 112), because the institutional practices are localized as well—in this case, in relation to their repeated destruction. Their stasis is not only shaped by direct assaults, however, but also is affected by the very constitution of knowledge landscapes.

In the absence of a fully recognized state of Palestine, political movements provide the primary support for memorials and national commemoration (Khalili 2007). Additionally, businesses (Bouillon 2004) and NGOs perform professional functions that otherwise might be under the purview of a national government. These include the production and preservation of empirical maps (Ibrahim 2011; Jensen, Abed, and Tellefsen 1997). Yet NGOs are often highly transitory, subject to outside control, and their future depends on fickle international aid cycles and grants (Hanafi and Tabar 2003).

This creates difficulties when conducting research, as Zureik, David Lyon, and Yasmeen Abu-Laban (2011, 1) point out in their recent study on information society in Palestine: "It is an understatement to admit that working in an emergent country like Palestine is not an easy task, particularly when the data … are hard to come by."[21] At the inception of the PPIB, the effects of Israeli military control combined with the prevalence of NGOs, at least partly as a result of the occupation, to ensure that the PA had no local map archive at its disposal.[22] Referring to the "almost total lack of basic data," in no small part because such data were "considered to be military secrets by the Israelis," one consultant simply states, "There were no maps" (Tesli 2008, 23).[23]

This enforced transitory character of Palestinian institutions interacts in complex ways with the stasis and mobility of the cartographers who have built and rebuilt those institutions. In addition to restricting the *scale* of their maps through the use of British colonial data, it shapes their ability to conduct surveys or otherwise collect data on the ground. This has the effect of restricting the *extent* of the new data that they can produce, with the end result that on PA maps, some areas are depicted in much greater detail than others. Attention to the multiple scales of PA mapmaking practices allows for a more detailed understanding of the specific ways that particular data sets are affected.

While discussions of Palestinian mobility tend to focus on local or regional movement (e.g., B'Tselem 2004), the mobility and stasis of PA cartographers and institutions must be viewed across geographic scales, including at international ones. The fact that many NGOs and PA ministries receive international aid with the precise aim of building stable state institutions led the noted Israeli journalist Amira Hass (2001) to observe that in its raid on the PCBS, the military was effectively destroying millions of euros of taxpayer money. Hass asks, Was the Israeli military, in citing security and intelligence concerns as the reason behind the raid, accusing Norway, Switzerland, and Germany of terrorism?[24]

Hass's question points to the ironies of PA internationalism, which can also be seen in the generally high international mobilities of Palestinian elites. Many of those I worked with had studied for advanced degrees internationally, such as in Canada, the Netherlands, Russia, and the United States, to name only a small sample. But the process of travel was often fraught with uncertainties. Besides the discrimination commonly found at checkpoints and airports, there is an ongoing prohibition against Palestinians operating their own planes without Israeli approval (which is not forthcoming) (PLO and State of Israel 1995). Moreover, the Israelis are attempting to build a settlement on the airport at Atarot, which would otherwise be the main Palestinian airport (Houk 2008).

To further add to these concerns, many Palestinians have no ability to obtain a passport from any nation. Permits and visas can easily be revoked—as in the case of Palestinian Fulbright fellows who were given grants but not the US visas that would have allowed them to take up the awards (Bronner 2008).[25] This is just one example of the numerous students and academics, among many others, who are systematically prevented from traveling from Gaza to the West Bank (Hass 2011a). The Israeli administration also imposes movement restrictions that target specific cartographers, such as the international travel ban on noted cartographer Khalil Tufakji, the former director of the Orient House Maps and Survey Department (MIFTAH 2010; interview 1, Palestinian cartographer and NGO director). Furthermore, in the broader context of forced international migration noted earlier, the ability to travel is not always a privilege. Many Palestinians must go abroad to find work or visit family members who themselves cannot enter the territories. So extensive international mobility can be a symptom of, rather than a remedy to, challenges to Palestinian stasis within the West Bank.

The relatively higher international mobility, for those Palestinians who can afford it, also stands in contrast to the well-known difficulties that Palestinians can experience when traveling within the West Bank.[26] While, as mentioned in chapter 1, the Gaza Strip has been virtually closed off to visitors since the rise of Hamas, the West Bank experiences periods of relative quiet where internal travel restrictions are eased somewhat and international tourism is common. At other times, extensive mobile checkpoints may be set up without warning, curfews imposed, and incursions or attacks may halt mobility almost entirely. The common practice of detaining Palestinians, including academics and technicians, further affects their mobility. Many who I spoke with noted that they regularly used their detailed knowledge of local streets and neighborhoods to help them find alternative routes in case of road closures.[27] Yet the very need for such knowledge indicates the difficulties imposed by the occupation forces.

These challenges also can be seen in the shifting locations of cartographic offices in the Jerusalem area and the resulting impact on their ability to update British data. PA offices are headquartered primarily in Ramallah, now the implicit future capital of a Palestinian state, and this is symbolic of the broader move to force Palestinians out of Jerusalem. Over the past decade, a variety of factors, among them the ongoing raids, have combined to push Palestinian organizations ever further from Jerusalem. Several of the NGOs I worked with were originally located there. Following the outbreak of the Second Intifada, however, Jerusalem was increasingly closed to Palestinians from the West Bank, and many cartographic NGOs moved east into the outskirts of the city so that all their employees could continue in their jobs. Once construction on the Separation Wall began, with the aim of cleaving the majority of the West Bank from Israel and its settlements, the NGOs moved again from the immediate vicinity of the Wall—which became the site of checkpoints and abandoned storefronts—and toward prominent West Bank cities like Ramallah and Bethlehem.

The move to the West Bank side of the Wall would influence the Palestinians' ability to collect information concerning those who remained on the Jerusalem side. For example, even while under the travel ban noted above, Tufakji continued to painstakingly rebuild a collection after the maps were confiscated, although he no longer advertised his location (interview 1, Palestinian cartographer and NGO director; Orient House 2001b). Given that each move would require a tremendous effort and outlay of funds, even those organizations that persisted under the occupation were made mobile in ways that rather than acting as a privilege, came to diminish their institutional stasis. The impacts of the specific forms of stasis and mobility experienced by the PA cartographers, both in institutional and operational terms, thereby serve to restrict the extent of the data that were produced. Yet this was not absolute. As I discuss below, the scope and geography of those restrictions themselves were quite varied.

In the Field, Stuck in the Car

Even after the 2008 handover of the PPIB to the relatively more stable MOP, the fact that their offices were ever on the move would serve to influence how geographic data were collected in the field. Cartographic field surveys are one major way that it might be possible to update British map data, and they involve using tools such as handheld GPS units—extended versions of the map applications on many mobile phones—to record precise locations for sometimes dozens or hundreds of buildings, roads, crops, and other features. Mobility is central to surveying, and it was one obvious factor that affected the PA's ability to collect data. Nonetheless, the cartographers I spoke with repeatedly mentioned that in many cases, the challenge was not simply accessing different areas. At one time or another, it was possible to reach most parts of the West Bank, with the obvious exception of the settlements and closed military zones (Ingham 2013; interview 7, PA transport cartographer; interview 8, PA transport engineer). Even so, they faced tremendous difficulties when attempting to stay in one place reliably without the intervention of settlers or the military. It takes time to observe and record data, and field surveys can require dozens of technicians, sometimes with heavy equipment. So in order to survey an area in the usual way, cartographers had to not only travel there but also remain. Thus stasis was as crucial to surveying as was mobility, or even more so.[28]

Yet reliable stasis still is not often achieved, especially given that as explained earlier, the West Bank is divided into the complex quilt of Areas A, B, and C (figure 4.2). Particularly in Areas B and C where the Israeli military has partial or full control, cartographers spoke of the efforts necessary to avoid drawing attention to themselves—and the Israeli soldiers still spontaneously appearing if the cartographers spent too much time in one location. Hanging around in the possession of expensive-looking instruments frequently aroused suspicion. As one cartographer pointed out, "We can move in Area C, but if we have equipment, it is very dangerous for us. If we have cameras or GPS, and if the soldiers stop us, they can arrest us. Two years ago, they arrested me and [my colleague] because they caught us in Area C. ... [We were taking] photos of demolished houses and GPS points" (interview 9, Palestinian NGO cartographer).

Many of those I worked with had been arrested at one time or another simply for carrying out their (legal and scientific) professional labors. Given that Areas A and B are surrounded by Area C, another pair of cartographers mentioned that although their survey was restricted to Areas A and B, nonetheless while traveling they often had to pass through Israeli checkpoints in Area C. They would need to hide their GPS units and cameras to avoid confiscation and summary arrest. The advent of pocket-sized GPS made their jobs easier, one engineer said, because they "look like mobile phones," although sometimes phones were also confiscated (interview 7, PA transport cartographer; interview 8, PA transport engineer).

For these reasons, in recent years PA and NGO cartographers have begun to develop the technique of mobile surveying, conducted fully within a car or taxi using handheld GPS. One mentioned that even though his groups work in Area C, "We are very careful. Like when we use GPS, we [take it] in the car and drive the car without anyone seeing anything" (interview 9, Palestinian NGO cartographer; interview 12, Palestinian civil engineer). Rather than visiting a single location one time to take a detailed reading, the use of cars allowed them to visit several times, for short periods, to take quick readings of different aspects of the landscape that could then be combined. So while their computer labs were gutted from within by the Israeli military, PA cartographers in the field were confined to their cars, which became circulating cartographic laboratories (Lee and LiPuma 2002). Their stasis and mobility thus were intertwined in ways that broke down divisions between the lab and field, and they strengthened Palestinian stasis not by rendering themselves immobile but instead by being constantly on the move.

There is a complex relationship between the extent of the area where data could be collected and the extent where cartographers were able to move and remain. For instance, even in a car, PA cartographers cannot conduct surveys throughout Area C, yet this does not mean that they never enter it, even if they don't intend to do so.

Although settlements and settler roads are well marked with signs as well as blocked by checkpoints, the boundaries between Areas A, B, and C are not always visible at ground level. The Israeli authorities only provide boundary information for the broadest-scale maps of the West Bank, so the internal boundaries are difficult for even most cartographic agencies to determine at finer scales. Cartographers might travel to Area C without realizing it (interview 5, Palestinian geography professor; interview 9, Palestinian NGO cartographer). Indeed, during my participant observation, at least one call per day would come in from landowners who wished to know whether their homes fell within Area A, B, or C, or some combination of them. Even if a home was on the border with only a backyard in an Israeli-controlled district, this was considered grounds for demolition by the Israeli authorities (ARIJ 2006a; interview 9, Palestinian NGO cartographer). Under such ambiguous circumstances, those I spoke with generally limit the amount of their data collection to Areas A and B alone, as much as it can be determined.

Measuring the Extent of the Occupation

Despite the limits on both their mobility and stasis, PA cartographers are still expected to provide updated maps of the "full" West Bank, including places like Israeli settlements that they cannot visit in person (interview 10, PA urban planner). This results in a situation where the ever-mutating stasis of PA cartographers in the field serves to reinforce the uncertain stasis of Palestinian towns and villages. By examining the grid that is often visible in the background of maps, the cycle can be seen (e.g., figure 4.1). In addition to aiding navigation, each square on the map is assumed to be depicted at the same level of detail as every other square. This creates tension in PA maps, however, given that Areas A, B, and C cut across different squares of the grid, and are surveyed in different levels of detail, but must be displayed together on a single map. The seamless transition between the data for each of the three areas obscures the lack of detail for Israeli-controlled Area C, which appears blank to the viewer, not as an omission on the map, but rather as empty space on the ground. This serves to erase dozens of rural towns and innumerable features that would be mapped with a consistent survey.

Such erasures can be self-perpetuating, because it is easier for Israeli authorities to justify moving or demolishing communities if they are even erased from Palestinian maps. The relegation of Areas B and C to the negative white space of PPIB maps is compounded by the limited extent of British cadastral maps, as shown in figure 4.5a. This may have contributed to the ultimate omission of Area C from the PA's jurisdiction. By using those maps, the PA cartographers were not only influenced by the rigid system of relatively broad (in comparison to the West Bank) scales in use by the

Mandate forces, as noted above. They also were affected by the fact that when the British did map higher-resolution data, this almost never occurred in the area that would become the West Bank. Indeed, figure 4.5b demonstrates the limited availability of Google Street View in the region, which suggests that even at present, data often are not publicly available for the West Bank.

So even as the data favored a broader scale, this was further compounded because the early colonial cartographers had concentrated almost entirely on more populated areas that, at the time, lay mainly along the coast. The British produced two main types of maps: topographic, with a focus on indicating elevation and land contours at a broader scale, and cadastral or real estate maps at much finer scales that primarily indicate the boundaries of individual parcels. The topographic maps were completed for the full British Mandate area, including the West Bank, by 1948, and their influence can clearly be seen in the maps produced by the PPIB (e.g., MOPIC 1998, 55–56). Yet the cadastral mapping was not completed for the hilly region that would become known as the West Bank, which at that time lay outside the extent of the survey. This was likely partly because the lands were not regularly the subject of disputes since the Jewish population in the area was minimal in comparison with their presence in population centers along the coast like Tel Aviv (Gavish 2005). It also reflects, though, how in the negotiations after 1948 that led to the formation of the West Bank and Gaza, the Palestinians were restricted to areas that were considered to be less valuable—and those were less likely to have been mapped in detail.[29]

The limited extent of the British maps thus are self-reinforcing in relation to the occupation's challenges to PA stasis. There is a cycle of omission. First, British cartographers concentrated on land that was determined to be economically valuable (to the British) and highly populated. Second, Palestinians were pushed to the West Bank—namely, land that was believed to be less valuable and less populated. Third, Palestinians are only given access to areas of high Palestinian population within the West Bank. Fourth, the lack of existing data for the less populated areas within the West Bank, and the movement and stasis restrictions placed on Palestinians, make it difficult for them to fill in data for areas that are blank on the map. Fifth, the fact that those areas appear as blank *even* on Palestinian maps is used to justify the demolition and seizure of the villages on those

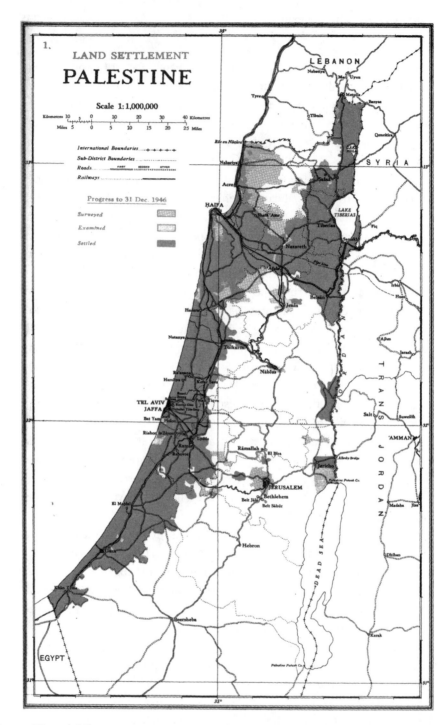

Figures 4.5a and 4.5b

At left, figure 4.5a shows the degree of the completion of British cadastral maps of historic Palestine as of 1945. At right, figure 4.5b is a screenshot that gives the extent of Google Street View that

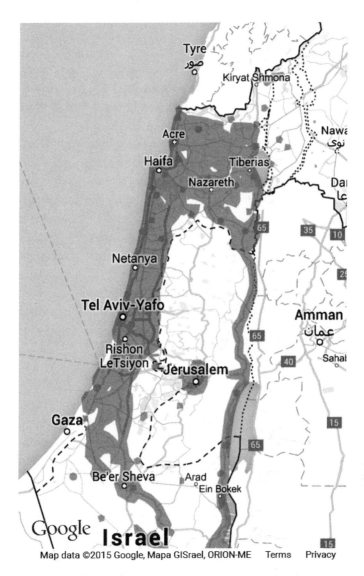

Map data ©2015 Google, Mapa GISrael, ORION-ME   Terms   Privacy

Figures 4.5a and 4.5b (continued)

is available in Palestine and Israel as of 2015. In figure 4.5a, the shading shows where the British had carried out finer-scale mapping. With the exception of a small selection of larger towns, the West Bank area was not cadastrally mapped (SOP 1948). In contrast, broad scale topographical maps, with less detail, were completed for much of the region shown, including most of what is now the West Bank. As indicated in figure 4.5b, neither the majority of the West Bank nor Gaza is presently included in Street View, but it does incorporate Israel-controlled East Jerusalem, specific Israeli highways in the West Bank (i.e., settler roads), and select Israeli settlements in the West Bank (Sheizaf 2012). The debate over the efforts to bring Street View's frontal-view style of mapping to Israel (as opposed to the more common bird's-eye view in many existing maps) is discussed in chapter 1. Figure 4.5a is taken from the Micha Granit Map Library, Faculty of Geography, Tel Aviv University, and used with permission.

(supposedly empty) lands. Sixth, Palestinians are prevented from accessing the demolished areas and therefore from documenting the destruction. This means they are less likely to be given control over these areas in years to come.

Through this circular process, Palestinians are pushed onto blank areas, then prevented from mapping the "blankest" portions of those areas, and then told that the unmapped areas can be seized precisely because they are blank. So the choice, in the Mandate period, to avoid mapping landowner- ship in the West Bank, combined with the ongoing Israeli occupation, means that the lesser-populated areas of the West Bank, including much of Area C, have simply been pushed into the white background of many pub- lic maps (e.g., figure 4.6). As such, the limited extent of colonial data is crucial, for areas that are left off maps become communities that are erased from the ground. This also helps ensure that their destruction will go undocumented—and in turn, that they will continue to go unmapped in the future.

Preservation in Pieces: A Diaspora of Palestinian Data

Despite the vast differences between the British and Israeli colonial regimes, they have combined to circumscribe the possibilities for the scale and extent of PA maps. Particularly in the early days of the PA, this resulted in broader, largely symbolic maps. The effects were felt strongly during the 1995–2008 period, although the process has continued after 2008. Still, the enduring stasis of colonial data has not only shaped PA efforts to build stasis for Palestinian institutions, but it also has affected the stasis and mobility of PA maps themselves.

For the stasis of PA maps tends to parallel the stasis of PA institutions. On the one hand, the PPIB maps were incorporated into a series of reports that were published, frequently in color, in multiple languages to assist in their circulation (MOPIC 1996, 1998). On the other hand, the fact that the reports were expected to plan for anticipated political realities—which in the aftermath of Oslo, included the predicted handover of Area C by 2005 (MOPIC 1998, 13)—meant that the durable, confident proclamations they contained were politically obsolete before they were even printed. This made the reports less useful and less likely to be preserved in the long run. The complete absence of a national Palestinian library, commonly part of

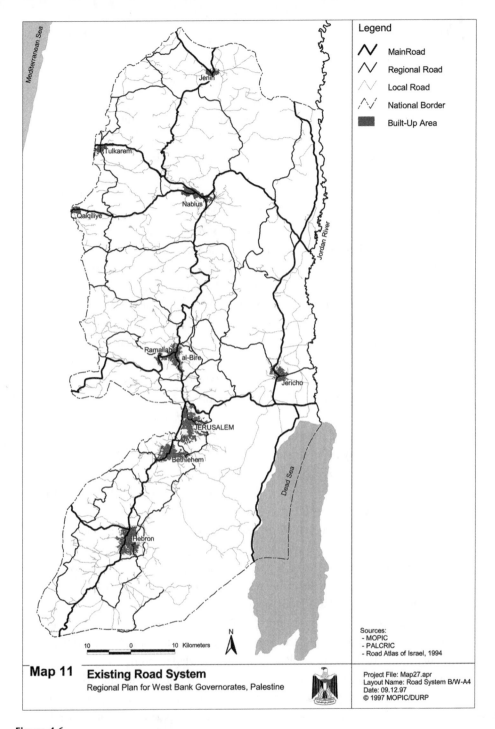

Figure 4.6

A PA map of major roads in the West Bank from 1997 (MOPIC 1998, 85–7). Although this map is based largely on data that originally came from British colonial maps, the information has been updated. Even so, the updates only appear for major cities in Area A, whereas inhabited areas in Areas B and C are not indicated—a result in part of the difficulties in conducting ground surveys due to the occupation (interview 6, former PPIB cartographer). A comparison of the overlap between the blank areas on this map and the background areas (Area C) in figure 4.2 can reveal the specific updates.

an established state, means that there is not yet any institution to collect these publications for longer preservation to ensure their long-term stasis, or send them overseas so that they are represented among international libraries' collections. The absence of such a library is not necessarily due to any lack of effort on the part of Palestinian officials but rather to the ongoing raids of the sort described earlier. In this context, documents are carefully preserved in PA offices such as the MOP and the local ministry in Hebron, NGO offices such as ARIJ and the Arab Studies Society, and the personal files of academics at universities such as Birzeit, Al-Quds, and An-Najah. Their very dispersal among a variety of local and international offices means that one particular confiscation, whether of an office or home, is less likely to wipe out all the PA publications. Yet it also means that their conditions and organization are decided at an individual or organizational level.

The commodification of data certainly plays a role as well in limiting the stasis of PA maps. Given the large amount of time and expertise that are required to produce data sets, even those adapted more or less directly from British maps, geographic data can be incredibly expensive. Valuable commodities are well guarded out of necessity in an area where the stability of institutions along with the mobility of goods and labor are highly restricted. Palestinian industry and private corporations started to develop in earnest only after the initial beginnings of the PA (Spinner 2012), and they did so in the context of a large influx of foreign aid that was input largely into NGOs and state-building projects like the PPIB. Although foreign aid has helped to set up the foundations for capitalist investment, for better or worse, the higher salaries and operational budgets of the NGOs have raised the cost of living, and through this the salaries for professionals overall. This makes things more difficult for new companies and helps to create an economic landscape that favors the international NGOs.

The dampened private economy results in a situation where NGOs and state endeavors are among the few producers or "factories" of economic value, often in the form of data or other knowledge products. This combined with the sometimes-uneven flow of international aid, due to political factors, means that data are held on to, for their economic value, against future uncertainty—uncertainty that is in part an intended result of occupation policies (Handel 2011).[30] The hoarding of data, however, reduces the number of maps that are made and makes the data more vulnerable to being misplaced, unused, or deleted, while also limiting the number of users. So the systematic effect of this local safeguard against uncertainty is to lower the level of stasis for cartographic data in Palestine overall.

In concert with the effects on stasis, the need to rely on British Mandate data has affected the PPIB maps' mobility, particularly in a situation where the state institutions that provide public data are both quite new and routinely challenged. Thus, one of the recurring issues that came up in interviews was the absence of reliable data sharing between PA ministries and Palestinian NGOs. While certain memorandums of understanding had been written, and some data were widely shared for specific purposes at least among PA ministries, the absence of a clearly defined hierarchy as well as competition among the PA and NGOs in combination with data's status as a commodity all served to restrict the unpaid exchange of data among different organizations. The PPIB itself was intended to collect geographic data from NGOs as opposed to creating them, I was told (interview 6, former PPIB cartographer), but the unwillingness of others to share data that had already been produced was one of the inspirations that lead the PPIB to embark on such an extensive digitization process (interview 6, former PPIB cartographer; Tesli 2008). On some level, it is to be expected that cartographers are concerned primarily with the stasis of their own institutions in circumstances where their work is regularly in question. However, it is important to delineate the ways that the occupation, in part through its control over ever-changing regimes of mobility and stasis, affects cartographers' daily lives in ways that feed into restrictions on data sharing more broadly.

Passages of PA Data

PPIB cartography was formed in landscapes that already had enforced *passages*—stable but "never-fixed" pathways of least resistance (Peters, Kloppenburg, and Wyatt 2010, 354–355)—for both the stasis and mobility of data. British maps were preserved, and digitizing them represented the easiest or most obvious passage for the PPIB cartographers to follow. Yet in order to build a national cartographic infrastructure, the PPIB cartographers had to redirect existing passages to build new forms of mobility and stasis for Palestinian cartographers, with the aim of increasing their control over their own maps.

The process of redirecting existing pathways served to influence the outcome of geographic knowledge production. This is acknowledged in the summary report of the PPIB, where the author (Tesli 2008) emphasizes the difficulties of producing new or alternative forms of data when the very fact of a Palestinian governing body making its own maps, in the late 1990s, was unique. Their ability to produce reliable maps despite the uncertainties, especially during the initial decade, when political conditions repeatedly were invoked to account for lost time (ibid.), is in part a testament to the degree to which PA cartographers have been able to collect and piece together British data, finding ways to preserve their collections in the face of repeated raids. As such, it demonstrates the theoretical and practical innovation, with respect to passages of mobility and stasis, that is involved in the material reconfigurations necessary to produce internationally recognized empirical knowledge. At the time, PA cartographers were sometimes criticized for being "unscientific," but such criticisms largely did not take into account the places in which the cartographers worked. Their ongoing forms of innovation thus have gone underappreciated, and will continue to do so unless scholars begin to take seriously the particular produced material landscapes of knowledge.

To a valuable extent, the PA was successful in developing new passages for knowledge in the West Bank.[31] The very limits on Palestinian cartographers' mobility and stasis, which led them to rely heavily on colonial maps, nonetheless also restricted their ability to control the scale and extent of their maps. The effects of the occupation, such that even PA government officials must be careful when collecting data in the field, served to reinforce the legacy of British Mandate maps. The process of pinpointing exact locations was founded on professional expertise and practical experiences of working through political uncertainties, military attacks, checkpoints, imprisonment, and confiscations, and the apparent seamlessness of PPIB maps (such as figure 4.6) belies the very complexity of the area under the jurisdictions in which they are produced. This was partly the outcome of the specific period of the Second Intifada together with the methods for surveillance and control that emerged in the context of the Israeli occupation at that time. Even so, it is still in effect in more stable periods.

Currently the differences between Areas A, B, and C are more pronounced than ever in terms of the geographic knowledges that are produced. If nothing else, by springboarding off the stasis of the British data, PPIB cartographers have pushed the landscape one step further toward accommodating passages that allow for greater Palestinian control over knowledge production in ways that are more ontologically and epistemologically diverse. Yet even as these passages are formed, their course is complexly shaped by the same colonial knowledges that they seek to overcome.

5 Validating Segregated Observers: Mapping West Bank Settlements from Without and Within

On the Appropriation of Empirical Facts

Scientific objectivity is built on the idea that two people in the same place, at the same time, can make similar observations. For science to work, it is necessary to believe that two people with similar training will be able to see the same thing (Daston and Galison 2007). This expectation has filtered into public discourse. That is why, when there are disputes, the opposing groups are often encouraged to physically come together to see for themselves. But that kind of collective observation may not achieve the desired result of encouraging consensus. For the most contentious debates concern not only different castings of the facts but also different facts. Such debates extend into the very constitution of the facts themselves. They do so in part because it is possible for two observers to be in the same position and observe the same object, yet see something different.

Rarely, if anywhere, are these differences more apparent than in debates over the scope of the Israeli settlement project in the West Bank. Ongoing series of competing reports about which buildings were built where, and when, serve to illustrate diverging observations and differing versions of the facts. Two sets of official reports on these facts are particularly illuminating. The first set began in October 2006, when the Israeli organization Peace Now published *Breaking the Law in the West Bank—One Violation Leads to Another*. The report consisted of an examination of Israeli settlements in the occupied West Bank. Through maps of property ownership, Dror Etkes and Hagit Ofran (2006) aimed to demonstrate that upward of 30 percent of the settlements were held on land that would be considered stolen—even according to the regulations of the Israeli government. The report was incriminating for the settlement project because it directly contradicted assertions that the Israeli West Bank settlements had been built in areas that were either unoccupied or owned by the state.

Many of the responses to the Peace Now report focused on the mathematical composition and scientific accuracy of the observational data used. As soon as the report was released, the Internet lit up with commentators who alternately praised and lambasted it, although the Israeli press was chastised for downplaying the implications of the findings, allegedly in the hopes of burying them (Benziman 2006). The intensity of the response only increased once an updated report was issued in March of the following year, inspiring further criticism (Peace Now 2007). "Peace Now's Blunder: Erred on Ma'ale Adumim Land by 15,900 Percent" was the headline of one reply (Safian 2007), which was made in response to changes in the Peace Now data between the two versions of the report.[1]

Yet rather than indicating a retraction of Peace Now's previous claims, as intimated in the reply, the updated report instead reflected newly available information. The data for the 2006 report came from a source inside the government who anonymously leaked it (Chadwick 2006). Then a few months after the report's initial publication, the Israeli government released an official data set to Peace Now, in response to the group's numerous Freedom of Information Act requests. Those official data were the source for the second report. For its part, the Israeli government issued no explanation as to why the new data differed from the previous set.

The changes between the two versions of the reports provide a cogent illustration of the unique position of Israeli NGOs in the landscapes of Palestine and Israel, and the ways that those occupied landscapes can shape the observations that are made. The "15,900 percent" critique assumes that Peace Now had complete freedom to move wherever it wished and use whichever data it wished. Yet this is seldom, if ever, the case. In this instance, Peace Now had specific types of access to specific types of data precisely because it is an Israeli organization operating within Israel—and one that therefore could submit Freedom of Information Act requests. As a result, its report was influenced by the constitution of the government source that it used, since the Israeli government had collected the data in order to serve its own needs. In addition, Peace Now's use of the official sources means that it was vulnerable to the government's ability to control and change that data, or release sets selectively, thereby causing the group to update its initial report.

Peace Now's position contrasts with that of a related Palestinian organization, ARIJ.[2] ARIJ produced the second set of reports that illustrates the ways that landscapes can shape the act of observation. For while Peace Now was critiqued in the context of its access to official data, ARIJ has been criticized precisely for conducting observations in areas where Jewish Israelis are not legally allowed to go. This type of response is evident in the reply to an online ARIJ report titled *Ecocide in [Taqu'a] Town*, which built on an earlier ARIJ (2004a, 2004b) publication.[3] The updated ARIJ report records one seemingly unobtrusive observation that received a particularly critical response. It states, "The Israeli Army randomly dispensed flyers on the farmers' lands informing [them] of the Army's intention to clear all trees (mostly olive) existing along the Israeli bypass roads" (2004a, n.p.). In his disparaging reply, Zev Wolfson of NGO Monitor claims to disprove the ARIJ report using methods of direct observation. Wolfson (2005) notes, "A visit to the area demonstrates that no trees have been uprooted from the area around Taqu'a."

By making this assertion, Wolfson appears to be heeding a call to go and see for himself. In so doing, however, he neglects the enforced differences among the (Palestinian and/or international) ARIJ observers and (Israeli and/or international) members of his own group. For each group of observers would have varying access to the site.[4] NGO Monitor's efforts to view the area from the Israeli road, which would be partially or fully unavailable to Palestinians, would differ from the view of ARIJ's workers from Palestinian farmers' fields. So Wolfson would find it particularly challenging to address ARIJ's (2004b) claim elsewhere in the same report that fields were burned and farmers chased away in areas—out of sight of the road—that might be largely inaccessible to Israeli settlers along with their supporters.[5]

Yet even if the NGO Monitor staff members were able to make timely visits to all the areas mentioned in the ARIJ report, it is quite possible that they might make varying observations based on their differing backgrounds and the changing forms of information at their disposal. Thus, for example, ARIJ would be able to draw on local Palestinian eyewitness accounts that would be all but unavailable to NGO Monitor staff members in the context of the Israeli occupation. So the ARIJ and NGO Monitor members most likely were not conducting their observations in the "same" place and time, and even if they were, they may not have made similar observations.

Where Are the Settlements Now?

The Peace Now and ARIJ reports are examples of the multifaceted ways that landscapes can affect the very constitution of knowledge. Peace Now was criticized because of its partial access to government data in Israel, while ARIJ was challenged for viewing facts from locations that most Israelis could not access in the segregated West Bank. Both responses draw on the presupposition that multiple observers can confirm observations. The facts are "the facts" because it is believed they are universally true—that they could be verified by anyone with the right training. But this characterization of the reports vastly underestimates the variegated constitution of scientific practice in geographic landscapes.

Far from simply revealing themselves as existing truths that are out there, waiting to be collected, facts must be laboriously *made* in particular places and times. Houses and people must be observed and counted, categories need to be devised and defined, and percentages and errors must be calculated. All this effort takes place in specific landscapes, through the arduous work of those steeped in particular scientific and technical traditions. Indeed, scientific practice has helped to produce the very idea of a separation between human and land, subject and object—or in the case of segregation, between subject and subject. The material and social consequences of these separations then feed back into technoscientific practices in those same landscapes.

That feedback and the diverging observations that result are the subject of this chapter. In what follows, I analyze several crucial ways that landscapes can and do shape the observations carried out during the long process of producing data and facts. To investigate how differing geographic positions might influence the act of observation—and through that, the production of facts themselves—I compare and contrast the maps of Israeli settlement expansion that are produced by these two organizations, ARIJ and Peace Now.[6] They are perhaps the two most prominent NGOs whose members map ongoing empirical changes in the landscape. Both NGOs share the goal of ending Israeli settlement expansion. Israeli settlements in the West Bank are considered illegal under international law. They have been constructed precisely to cement the hold of the Israeli government on areas believed destined for a future Palestinian state. However, to date there is little agreement on the extent and location of these settlements. Rather than using agreed-on facts to support political arguments, actors often debate the intrinsic composition of the facts them-selves, including the ways they are made. As a result, calls to use the facts to settle disagreements or confirm facts through further observation rarely lead to consensus.[7]

Thus, the challenges of validating observations in the context of the occupation stem from the ways that such fact making is firmly and some-times oppositionally rooted in social worlds born out of experience (Radder 2006). The advance of the settler movement along with the resistance that results has led the occupation forces to impose extreme forms of segrega-tion in the West Bank and East Jerusalem—a segregation that aims to con-sistently privilege Israelis.[8] Jewish Israelis are formally allowed only *within* the settlements, whereas most Palestinians with West Bank visas can only travel *within* Palestinian areas that lie outside the settlement fences, walls, and other barriers. Similarly, Palestinians are increasingly prevented from entering settlements, viewing them from *without*, while Jewish Israelis pri-marily view Palestinian areas from *without*. This double, thoroughly unequal segregation fosters diverging sets of experiences among Palestinians, Israe-lis, and internationals. It can result in sometimes widely differing observa-tions of places and events, even when people occupy spaces that are allegedly the same.[9] Segregation and its related power asymmetries there-fore have complex consequences in terms of shaping empirical knowledge of the occupation.[10]

A View of Har Homa, the View from Har Homa

The positions of ARIJ and Peace Now within the landscapes of the West Bank also reflect power imbalances between Palestinians and Israelis, both locally and internationally.[11] In contrast to the broader reception of Israeli accounts, until recently Palestinian perspectives were not widely represented in Europe and North America. Even on the occasions when Palestinian intellectuals and activists discussed the conflict with a (non-Arabic-speaking) international audience, their experiences were frequently downplayed, in part because they didn't fit the dominant conception of what a *voice* would need to be in order to be considered credible (Spivak 1988, 1999). As a result, their accounts were sometimes characterized as unscientific.[12] They were criticized for providing not facts—meaning, not statistical facts—but instead stories and witness testimonials. Palestinian cartographic efforts are in part an attempt to respond to such accusations by adopting cutting-edge statistical and cartographic visualization techniques.

The international perception of Palestinians is more than a minor matter, because international public opinion has long played a central role in the Israel occupation. Lori Allen has remarked that "the problem of how to make themselves audible and visible has been a central stumbling block" for Palestinians. Allen (2013, 35) points out how in the mid-twentieth century, "Palestinians were without any territory or institutional platform to express their national aspirations." This political and geographical context made "establishing the credibility of the testimony takers just as important as that of the testimony" (ibid., 50). In contrast, although anti-Semitism continues to significantly affect the broader reception of work by Jewish researchers, Israeli scientists have been successful overall in terms of presenting themselves as credible observers. This is in no small part a result of the legitimacy afforded by the state of Israel as well as the perception of Israelis as culturally European and racially white (see chapters 2 and 3).[13]

These power dynamics have crucial implications in terms of the international validation of Palestinian eyewitness accounts of the occupation. Validation depends on the ability to reproduce observations, yet observations of the effects of the occupation are inaccessible for many outsiders. So before delving further into the international reception of Palestinian data, it is worth outlining the steps of this *invalidation* of Palestinians' experiences in greater detail. First, Palestinians are targeted as a group under the occupation, so they experience particular injustices precisely *because* they are identified as Palestinians. Second, by definition, those aspects of the occupation are not directly accessible to non-Palestinians who can't or are far less likely to personally experience them. But third, this means that in order for those unique experiences of occupation to be transformed into international knowledge, the outsiders must acknowledge Palestinians' accounts as valid observations of events. Fourth, however, such an admission is absent precisely because of the same discriminatory tendencies that make it possible to target Palestinians in the first place, thereby bringing us back to step one.

Such an acknowledgment of the possible validity of Palestinian accounts is exactly what has long been lacking in international debates. On an international level, Palestinian advocates risk speaking a private language

(Wittgenstein 2001), and this presents another level of power imbalances. A hypothetical scenario illustrates the common response to Palestinian claims. Imagine that in the course of negotiations over the occupation, the Palestinian team told the international negotiators, "The Israeli military demolished our homes." The negotiators might then reply, "Are you sure? Because all our homes are still standing."[14] To the negotiators, it might seem obvious that demolitions are rare, but that's only because their homes wouldn't be targeted *precisely* because they're international negotiators who might have the power to retaliate. To counter such tendencies, Palestinian advocates must not only produce facts. They must develop a community of Palestinian observers, whose accounts are verifiable within the community, as part of a legitimizing process that in turn can convince international organizations. To do this, there is a need to develop an institutional infrastructure of validation that allows for observations to be verified among professionals from within the group, while simultaneously presenting their observations in a statistical form that is internationally recognizable.[15]

The power imbalances that sometimes preclude validation are not just confined to political negotiations. They also condition the reception of scholarly research both locally and internationally. To offer an example, Moshe Brawer, one of the founders of geography as a discipline in Israel, draws on Orientalist stereotypes of inconsistent, irrational Palestinians (Elia 2004, 2005) in his review of the edited atlas by Walid Khalidi (1992), a highly esteemed Palestinian geographer working in the United Kingdom.[16] Brawer (1994, 337–338), whose work is equally respected, places particular emphasis on his criticisms of Khalidi's fieldwork method, contending that it "leaves so much to be desired—in systematic and consistent description of geographical features, in accuracy, and in scholarly observations."[17] What Brawer does not mention, though, is how Khalidi's position as a Palestinian academic might lead to challenges in his access to sources and field sites—and thus to potential difficulties conducting the types of extensive, standardized observations that Brawer expects.

These portrayals also affect research that is completed in government or NGO capacities. Reacting to negative characterizations of their work, one cartographer in a Palestinian NGO related that "the first thing when you go to conferences is, 'What's your sources?' ... In the early days, it was all about emotions ... but soon enough they ask you, 'What's your source?' ... In the early days, we'd say, 'What source? I'm telling you,' and they would go, 'It's your word against theirs. What facts do you have?'" (interview 11, Palestinian NGO cartographer). His words point to a twofold labor on the part of Palestinian organizations and academics. On the one hand, it is necessary to be recognized as a credible observer in a political game that depends on such recognition (Markell 2003). On the other hand, it is necessary to formulate claims precisely in a way that will be heard—to produce a community capable of putting forward the allegedly objective facts the serve as a metaphoric entrance fee to the debates.[18]

Throughout its history, ARIJ has worked to establish just such a community for international validation. Initially started as an environmental group to further sustainable development methods for Palestinian agriculture, ARIJ has become a major producer of civil maps for use in local as well as international debates. Founded in Jerusalem in 1990, ARIJ predates the advent of the PA. As such, it was one of the first organizations in the Palestinian Territories to begin making digital maps using GIS software.

Jad Isaac, who comes from Bethlehem and holds a PhD in agriculture from East Anglia University, has headed ARIJ since its inception. Currently, ARIJ includes over fifty employees, with the Geo-Informatics Department (the primary GIS unit), and Urbanization Monitoring Department (which includes settlement monitoring) together employing ten to twelve of those staff members. Funded in part by international donors, including the European Union, ARIJ nonetheless primarily employs Palestinian cartographers from a variety of religious and ancestral backgrounds, mostly from the West Bank but also from East Jerusalem, not to mention a revolving group of international fieldworkers like myself.

As an organization that combines nationalism and empirical methods, Peace Now functions as an Israeli counterpart to ARIJ. A pro-peace Zionist organization, Peace Now emerged as a mass movement in 1982, after Israel invaded Lebanon with the aim of expelling the PLO, which was then stationed there. When the nonviolent First Intifada began in 1987, Peace Now came out in favor of negotiations and backed the 1993 Oslo Accords, although its support base decreased during the Second Intifada, once violence escalated. But it continues to work for an end to the occupation as a strategy to enable a viable Palestinian state while, in its view, strengthening Israel.

Currently headquartered in Jerusalem with about twelve employees, Peace Now is one of the largest and most widely known Israeli peace organizations. Together with its US sister organization, Americans for Peace Now, Peace Now's advocates have undertaken one of the longest-lasting and expansive settlement mapping projects. Its work builds on earlier geographic surveys of the West Bank by Israeli NGOs, such as that conducted by Meron Benvenisti and Shlomo Khayat (1988). Begun in 2002 by Etkes, the Settlement Watch division is the main cartographic arm of the organization, and Ofran has directed it since 2007. It was with Ofran that I first entered a West Bank settlement, where I was dumbstruck by the sheer suburban banality that lay beyond the electrified fences and evacuated hills.

The Power of Observation

Valérie November, Eduardo Camacho-Hübner, and Bruno Latour (2010, 585–587) have described how digital cartographic technology further breaks down the assumed divide between the material world and map by bringing ever-greater numbers of maps outside on phones and GPS devices, while simultaneously pulling reams of data traces into the lab. Here I build on this insight to analyze how produced material dichotomies, such as the division between Palestinian and Israeli areas in the West Bank, nonetheless can be built back into the map.[20] My approach has similar concerns as those in related fields of the study of technology, such as Eyal Weizman's (2011, 2014) work on forensics. To this research, it contributes an awareness both of the obduracy of colonial legacies, among other material and social configurations, and pervasiveness of the intersectional relations of power and privilege based on gender, race, ancestry, region, ability, and economic class. I therefore call for greater focus on how material and social forces feed back into even the most critical and reflexive academic work, thereby challenging efforts to export critical methodologies across diverse regions and groups.

In what follows, I first examine the geographic implications of feminist standpoint theory, told through the cartographic concept of triangulation. Standpoint theory and STS studies of practice have much to contribute to each other. In the reformulated version presented in this chapter, standpoint theory provides an important locus for thinking critically about the relationality and materiality of difference, while STS research allows for a fuller exploration of conceptions of objectivity, including the role of privilege and difference in the laborious production of facts. The combination of both allows for an analysis and rethinking of the role of landscapes in technoscience, which is here accomplished in light of the practice of *triangulation*.

Triangulation is a surveying method where observations of known points in the landscape are used to determine the coordinates of an unknown point. Here I use triangulation in a more metaphoric sense—namely, to situate the account offered in this chapter in relation to the two accounts, one Palestinian and one Israeli, that are analyzed. Second, I turn to the settlements, using triangulation to analyze how ARIJ's position without and Peace Now's position within Israeli areas in the West Bank in turn shape each NGO's resulting maps. Ofran travels regularly on fact-finding trips within the settlements, and this, combined with her increased access to data within Israel, affects the maps that she produces for Peace Now. However, this positioning has nuanced effects, particularly in terms of the display of Israeli military bases and related changes in settlement boundaries, which I recount using maps of the settlement industrial area Mishor Adumim.

Third, I turn to an analysis of Palestinian communities in the West Bank in order to investigate how ARIJ's position viewing Palestinian areas from within and Peace Now's position viewing them from without differently shape the cartographic methods that they use. In order to get at the complexity of the power imbalances between the two organizations, I analyze ARIJ's land use / land cover (LULC) cartography, which seeks to fill in the Palestinian areas that have long remained blank on many maps of the West Bank. I explore how ARIJ cartographers' ability to adapt international LULC classification systems to fit the features around them depends both on their scientific expertise and everyday experiences as Palestinians in the unequal landscapes of the West Bank. Furthermore, through a network of Palestinian observers in the West Bank, Gaza Strip and beyond, ARIJ often obtains early knowledge of the military orders for land confiscation that are distributed to Palestinian farmers. In contrast, not only are Peace Now fieldworkers' maps of Palestinian areas less detailed than those at ARIJ, but their position viewing Palestinian areas from without divorces them from communities of cartographers, like ARIJ, that might otherwise be potential contributors to their efforts.

In the conclusion, I return to the notion of reflexivity, but do so through a call to refractive analyses. As I use it here, *refractivity* refers to the practice and theory of reflexivity when it is used specifically in reference to the power asymmetries of spatial and material situatedness. The practice of refraction thus requires ongoing efforts to foster an embodied awareness of relational and grounded positionalities, including those of individuals, groups, disciplines, theories, and practices. But first I turn to triangulation, which provides a method for thinking through ways to contextualize international knowledge about the occupation.

Triangulating Standpoints: Observations Are Made in Landscapes

Observations of landscapes are themselves made *in* landscapes. This point is not lost on cartographic surveyors. So it is with some irony that triangulation, once the dominant scientific method for surveying and mapping, has since been claimed by feminist studies of science in a way that de-emphasizes its geographic influence. In cartography, triangulation is a way of viewing other points on the ground specifically in order to gain a better understanding of one's own position. To triangulate a point whose exact coordinates are unknown, a surveyor stands on that point and sights other known points through an instrument that might look something like a small telescope affixed to a tripod. For each point viewed, the instrument gives a reading of the angles between that point and its own location. Once two known points have been sighted, the surveyor can combine the angles with the coordinates for each of those points in order to calculate the coordinates of the unknown position where they are standing.[21] Triangulation could be used to extend a grid from mapped areas into unmapped ones. Pushing forward, the surveyors could determine each new point by sighting points behind, in the mapped area. This provides an interesting geographic counterpoint to philosopher Walter Benjamin's ([1974] 2001) angel of history, who is driven into the future by a "storm of progress" while facing backward onto the debris of the past. Surveyors often faced backward, sighting mapped territory through their instruments as they pushed ever further into areas that they viewed as frontiers, as terra incognita.

Triangulation likewise has colonial roots, and its use by the British in historic Palestine is one of the main reasons why British colonial maps of the region are considered so effective (Gavish 2005). Yet this leapfrogging over known points into "new" areas was not haphazard. It also relied on a totalizing imagination of a coordinate grid that potentially could cover the entire globe. Regarding nineteenth-century British triangulation efforts in Egypt, Timothy Mitchell has noted that the production of the nation was predicated on an attempt to view the land in its entirety. This could equally apply to the British colonial surveys in Palestine (figure 5.1):

The survey began by establishing coordinates not within the village but across the entire country. ... The nation was emerging as this space, this material/ structural extension, within which villages, persons, liabilities and exchanges could be organized and contained. ... The connections, linkages, commands, and flows of information that made up this political order ... appeared to arise in the space of separation between the land and the map. (Mitchell 2002, 90)

So in the early and mid-twentieth century, first the British and then the Israelis attempted to make a nation out of historic Palestine by imposing a separation between the land and map, and viewing the region as one definable territory.[22] Triangulation was a crucial technique that made the totalizing separation possible.

In this context, it is perhaps surprising that the notion of triangulation, which seeks to abstract a particular view within a landscape into an abstract network, has been adopted by social scientists writing on Palestine and Israel who seek to make abstractions more concrete by demonstrating the intimate connections between abstraction and material positioning.

Figure 5.1

A map of the British triangulation network in historic Palestine in the late 1940s (SOP 1948). Every point where the lines intersect is a node in the network. Several known points in that network would be temporarily marked on the ground using flags, beacons, stone towers, or simply by having someone stand in that location. An observer would then visit the approximate location of one unknown point and use a surveying instrument to view and measure (from afar) the markers at two to three of the nearby marked nodes. Calculations based on those measurements then allowed surveyors to determine the precise location of the unknown point where the observer was standing. Once this was achieved, the markers would be moved forward, and the observer could move to a new unknown point and begin measuring anew. In this way, surveyors could determine the latitude and longitude of every point in the network, connecting the dots to plot and map the overall territory. Taken from the Micha Granit Map Library, Faculty of Geography, Tel Aviv University, and used with permission.

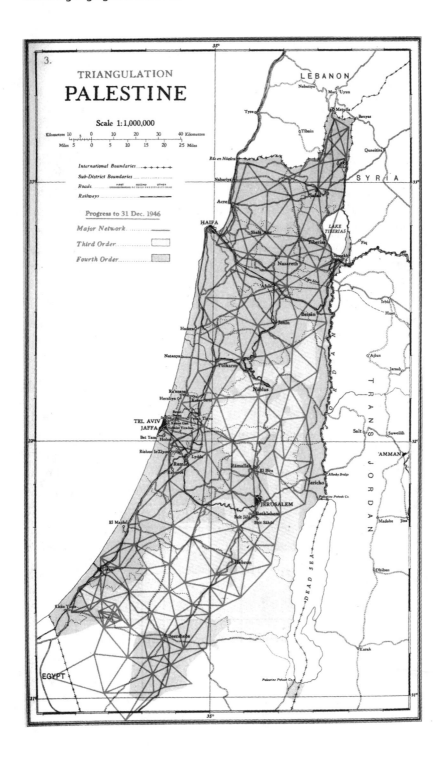

Wendy Pullan (2013, 125), for instance, speaks of "extreme binary vision" in segregated Jerusalem, and Michael Fischer (2006, 162) explicitly argues for "two-eyed" narratives that rest equally on Palestinian and Israeli accounts.[23]

The metaphoric aspects of triangulation have not escaped social theory more broadly, where it has become used in quite a different sense. In feminist studies of science and contemporary social sciences more generally, triangulation is employed in two related ways. First, several methods of observing are combined, usually including both qualitative and quantitative approaches, with the idea that the methods will complement each other (Harvey and MacDonald 1993; Webb et al. 1966). Second, different data types and sources are brought together with the aim of forming a more textured view of one phenomenon (Denzin 1970, 1977). Scholars such as Sandra Harding (1998, 2008) have worked to triangulate different perspectives on science, thereby combining several points of view, particularly from traditionally marginalized groups. In sociology, Greg Smith (2006, 114) has explored how Erving Goffman advocated a combination of "naturalistic observation" and interviewing. Goffman claimed that in conducting fieldwork, he tried "to triangulate what people are saying with events" that he observed (Goffman 1989, 131; Smith 2006, 114).

On the one hand, these important social science uses of triangulation provide a working strategy for combining diverging accounts and perspectives. On the other hand, too little attention is paid to the geographical Cartesianism implicit in such use. In order for practices like triangulation to be thoroughly transformed for critical and relational theory, it thus is necessary to conduct a more thorough examination of its cartographic and colonial roots. As a method for developing a Cartesian grid over an entire region, for example, cartographic triangulation furthers a specific notion of the *local* scale—namely, as a particularized scale that is always already related to an overall macro-scale grid. So calls to compare particularized, local methods through (social science) triangulation may just be ushering in totalizing macro notions through the back door. Certainly, particularism is not the only way to conceptualize the local. Michael Lynch (1993, 125) defines the *local* as a term that "has little to do with subjectivity, perspectival viewpoints, particular interests, or small acts in restricted place. Instead, it refers to the heterogeneous grammars of activity through which familiar social objects are constituted."

Yet the identification of the local with particularized knowledge, as Lynch rightly notes, has a long history that perhaps cannot be disavowed so easily. Indeed, the particular becomes relevant precisely in the context of the development of an opposition between the local and the global as well as the broad scales and coordinate grids that were imposed through practices like triangulation. Thus, ethnographic research on the local, including "small acts" and "perspectival viewpoints," have become important precisely in relation to the continued dominance of macro scales rather than irrespective of them. The geographic context of triangulation is significant precisely because it shows all the effort required to bring a (narrowly particularized) Cartesian version of the local into being.

Triangulating Standpoint Theories

In order to demonstrate how such links both persist and mutate, it is necessary to reconceive of triangulation as an ongoing, performative, and relational act. Instead of attempting to make a "true" observation, it can be rethought as the study of how relationships between observers can serve to reconfigure the observers themselves, in addition to their observations and inscriptions (Latour 1986). Perhaps the most thorough effort at reclaiming triangulation has taken place in standpoint theory, a body of research that attempts to rework the values of international science in order to take greater account of how scientists' identities inform the ways that they practice science (Harding 1998).

Standpoint theory allows for an examination of the gendered and racialized contexts of triangulation, which in concert with its geographic and colonial roots, also intrinsically affect the constitution of knowledge. For regimes of sexism and Orientalism long identified women and people of color with the particular and local, on the one hand, and men and white people with the universal and global, on the other. But standpoint theory work on gender, race, and observation has drawn attention to the links between observation and control, specifically in terms of white masculine "mastery" over feminized, colonized land (Rose 1992).

Standpoint theory has been controversial since its inception, attacked, on the one hand, during the "science wars" for being too radical in its reconfiguration, and on the other hand, for being too Eurocentric and essentialist, for backhandedly upholding dominant scientific methodology (Harding 2004b; 2006, 92–93). Given its colonial past, the role of

Eurocentrism is certainly pertinent to a fuller understanding of triangulation.[24] Indeed, even standpoint theory's most nuanced and contextualized attempts to coordinate observations and identities might still imply that there exists some utopic future where all those observations would or should align. Attributing observational differences solely to injustice has the danger of implying that by nature, people are the same. So standpoint theorists run the risk of implicitly suggesting that if all injustice were to cease, then everyone simply would agree. Yet all methods bring concerns, and this tendency, to attribute all difference to injustice, is best something to be addressed in specific cases as opposed to one that can be corrected outright. Furthermore, in the debates' focus on standpoint theory's deficiencies, two of its contributions have been too easily dismissed or overlooked.

First, standpoint theory provides a method for democratizing science for those concerned with incorporating textured conceptions of identity, in relation to structural inequality, into the heart of scientific practice. Standpoint theorists, second and most relevant here, rightly focus on the role of subject formation in the study of scientific practice. Namely, they actively reconceptualize the link between the points of view among different social groups at a variety of scales and scientists as subjects themselves. They therefore draw attention to the relationships, including triangulations, between knowledge production across multiple scales. At the same time, however, they productively diverge from the more problematic assumptions of cartographic triangulation, where the subjectivity of the cartographer should be effaced in the quest to produce objective knowledge. Standpoint theory thereby cultivates a thorough respect for the at times unpredictable and unexpected ways that alterity permeates the practice of science and identities of scientists. Through its relational understanding of triangulation, this chapter forms part of this reclaiming of triangulation. As such, it is in line with Nira Yuval-Davis's call (2012, 49) to better understand "the transitions from positionings to practices, practices to standpoints."[25]

Yet, while this chapter draws on the work of Harding (1998, 2004a, 2006, 2008), it is something of an inversion of the main goals of standpoint theory. For whereas standpoint theorists seek to involve individuals from standpoints across power imbalances in the production of science, here I investigate how a particular standpoint is produced through the practice of science, and therefore how that standpoint is coconstitutive with scientific practice. So I look not only at how differences between scientists shape their practices but also at how segregated landscapes help produce different scientists. Furthermore, while Harding ends by reaffirming objectivity, albeit an objectivity that incorporates social and political values into the heart of technoscientific practice, in contrast, I analyze objectivity as a practice with unique historical and geographic positions (Daston and Galison 2007). In the process, I also draw on Donna Haraway's related conception of *situated knowledge* (1988).

By comparing and contrasting the work of ARIJ and Peace Now, I thereby triangulate my own positioning into the analysis. I serve as an international observer whose presence represents an intervention (albeit not necessarily a positive one), and yet one whose analysis and methods *also* might be transformed through the relationships that are developed over the course of the triangulation. This chapter as a whole, then, consists of a reclaimed form of triangulation, but one whose three perspectives neither consistently overlap nor diverge, and don't necessarily occupy commensurable spaces. I move away from the sense of triangulation as a means of eliding difference, instead attempting to formulate dialogic relationships, but with attention to related asymmetries of power. In addition, while building on standpoint theory, I examine the spatial aspects of triangulation's history from a critical perspective. Rather than combining points to calculate one true set of coordinates, I take the differences among points and center them as the main subject of the analysis, neither obscuring them nor treating them as absolutes, straddling between speech and listening, involvement and waiting, dialogue and an awareness of the limits of my own understanding.

This modified use of triangulation moves away from the notion of *position* as a fixed point or set of coordinates, and away from a reduction of location to position. Instead, it gravitates toward an understanding of how any single location or subjectivity stands in dialogic relation to a variety of landscapes.[26] So I both employ a recast notion of triangulation and seek to cast social science triangulation itself in a new geographic light. With this in mind, I now turn to an analysis of the Israeli settlements as sites of geographic research. To begin, I specifically analyze how Peace Now's ability to work within the settlements and ARIJ's position viewing them from without affect the settlement maps produced by both groups.

Mapping Israeli Settlements: An Atlas of Dirt

"You see them?" Ofran asks. "Yeah. The dirt mounds?" "I didn't see them before, but now I notice." As she speaks, she accelerates her white SUV up the Israeli road outside the Israeli settlement of Tekoa, which lies in the West Bank midway between Hebron and Jerusalem, strategically close to the Palestinian village of Taqu'a discussed at the beginning of this chapter. Ofran has let me tag along as she takes an international couple on a field tour into the West Bank. In addition to pointing out significant sights and events, occasionally she stops the vehicle to take photographs or ask questions of those we meet.

As we wind our way in and out of the settlement networks, dodging the occasional excavator or dump truck parked along the road, Ofran points out the different types and ages of the dirt mounds that the construction teams have dredged up from the earth. There are a staggering variety, from waste that will be gone in a few days, to dirt piles topped with grass that have likely been there for months or years, to the plateaued hills that are the future sites of settlement apartment blocks, grocery stores, synagogues, and swimming pools.

Israeli settlements in the West Bank are built and expanded in chunks, as multistory developments that are imposed on the land. They require an excessive reshaping of the terrain before they can be built. Dirt is piled up in some areas, and valleys are carved out in others. Concrete retaining walls are sunk in between the artificial valleys and plateaus, while roads encircle the newly reshaped hilltops. For example, figure 5.2b (right) is a typical photo taken on one of the Settlement Watch's tours of the West Bank. It illustrates the size and extent of the digging into one such hill to the east of Jerusalem. Figure 5.2a (left) is another photo from a field visit that shows the roadblocks set up by the Israeli military to slow the movement of Palestinians even in formally annexed areas of Israel like East Jerusalem.[27]

Field visits like these are a central part of digital cartography. Aerial and satellite images—the types of images seen in Google Earth or Google Maps—are sometimes used in place of fieldwork. Due to costs and legal requirements, however, such photos are out of reach of many NGOs that seek to make publicly available settlement maps. As a result, both in Palestine and Israel, and beyond, fieldwork is a regular part of the process of making a map, even in cases where aerial photos are the primary source of data.

There are many reasons to conduct fieldwork. The types of ongoing changes that are typical of settlement expansion are too nuanced and rapid to be readily legible on contemporary aerial photographs. Updated images are too expensive for many NGOs to purchase for a time span shorter than several months, if not years, in sufficient quantities to map the scale of the West Bank—a scale that is necessary because the settlements are purposefully scattered throughout the region. When purchasing images, it is rare to obtain one of the entire West Bank. Instead, organizations regularly buy specific images of a few selected sights at high resolution, and the choice of which updates are necessary would also be based on fieldwork. In addition, even in cases where high-resolution aerial images are available, not all features are visible from above. There are numerous ways that particular changes might be interpreted.

So aerial photographs have not replaced fieldwork, which continues to be viewed as an essential component of digital cartography (interview 20, Israeli NGO researcher). Indeed, one cartographer for B'Tselem, an Israeli human rights organization, mentioned the benefit of having a fieldworker suggest updates to his maps because he works primarily from a computer. Paraphrasing the fieldworker's corrections, he said, "It's not like this. This is here, and this is here, and this is here." Describing the types of updates she wanted him to make, he explained, "She'd say: 'This checkpoint doesn't exist. The wall here is built, but you say it is only planned, and there is a road here that I've driven many times, but it's not on your map" (interview 21, freelance Israeli NGO cartographer).

Even so, the situation is different for Palestinian and Israeli NGOs. With the exception of the images they take from distant hills with high-resolution cameras (interview 9, Palestinian NGO cartographer), Palestinian cartographers sometimes must rely on commercial satellite imagery and remotely sensed data because of restrictions on cartographers' mobility. In 2003, Majed Abu Kubi (2003, 71), who was then working at ARIJ, noted that he turned toward satellite data owing to "the fact that we don't have access to all parts of the West Bank to obtain ground truth points and [also because of] restrictions on obtaining aerial photos ... due to security reasons." Such restrictions were particularly evident in the first years of the twenty-first century, yet they are still very much in effect.

Figures 5.2a and 5.2b
Blocking roads and building roads. Peace Now's Settlement Watch took the photo on
the left (figure 5.2a). The picture is rather unique because although it was recorded
by an Israeli organization, it shows a roadblock in the Palestinian neighborhood of
al-Issawiya from within. This is only possible because the neighborhood is in Israeli
occupied East Jerusalem. Israelis, limited to Israeli areas in the West Bank, are legally
allowed to travel to East Jerusalem. Figure 5.2b (right), a photo also by the Peace
Now's Settlement Watch, shows the E-1 settlement area from within, recording the
stages of road and infrastructure construction (Friedman 2012). Both images used
with permission.

Ofran's position is relatively unique. Through images like the one in
figure 5.2b, she has developed an extensive photographic record of settlement construction, and with it a method for distinguishing between different dirt and rock piles. She travels the settlements often enough to be able
to tell which piles are new, which are bigger than before, and occasionally,
which have been abandoned. The ability to go to such areas and photograph the views from within is a core part of the work of Ofran, and Peace
Now more broadly, as they document what happens inside the perimeter
that goes up around each swath of land that is confiscated, whether for

Figures 5.2a and 5.2b (continued)

military purposes, settlement building, or both. Thus, her travel in the settlements affords a singular opportunity to collect data for Peace Now—one that is denied to the researchers at ARIJ.

Settlements from within: Maps without Military Bases

As an Israeli citizen, Ofran's ability to enter the settlements is part and parcel of her ability to enter Israel. Yet her status as a Jewish citizen of Israel also conveys multiple further benefits on her cartographic efforts. The fact that Ofran can travel within Israel is significant, too, as it gives her resources that allow Peace Now and other Israeli NGOs to petition the Israeli government for official data, like those used in the reports discussed earlier—a process that can take years (interview 14, Israeli NGO director; interview 19, Israeli NGO urban planner).

In addition to allowing access to the settlements, their mobility within Israel means that Peace Now has greater access to unofficial aerial photographs than does ARIJ. Several times per year, Peace Now is able to charter small planes to take its own aerial images—which although expensive, are still cheaper than buying commercial photographs, with the added benefit that the images are not under copyright. During these trips, Ofran sits at the window of a small plane with her digital camera, snapping photos of the landscape below.[28] This is not to say that Ofran has unrestricted access, however; far from it. Sometimes her reputation precedes her, and she may be refused entry, although by Israeli law she should be able to visit all the settlements (e.g., Klibanoff 2009).[29] Nonetheless, her ability to travel within and above both Israel and the Israeli West Bank settlements thoroughly shapes her maps.

The information that Ofran gathers within the settlements allows for a much finer level of detail and knowledge than would otherwise be possible, such as changes in the locations and size of specific buildings.[30] But Peace Now's position also affects the very constitution of the data that their cartographers make and use. This can be seen in Peace Now's treatment of one crucial area type that proliferates throughout the West Bank: Israeli military bases. Sensitive sites are routinely censored by law from aerial images released by private Israeli firms. Cartographers have numerous examples of images where a particular detail—for instance, a sandy hill or tuft of scrub brush—has been copied and pasted over another area in the same image to hide a military installation. So one image might contain two identical dirt hills, but one of these hills would cover a series of military buildings and roads (compare figures 5.4, 5.5a, and 5.5b)—prominent features that might be quite visible through a car window (interview 1, Palestinian cartographer and NGO director; interview 11, Palestinian NGO cartographer).[31] Yet even when uncensored high-resolution satellite photos are available, they need to be standardized and transformed in order to use them for maps (figure 5.3). Detailed images can cover only a small area at a particular time. Different photographs may be flattened or projected in different ways. This causes significant discrepancies between images, and can require ongoing and laborious technical adjustment when, as in most cases, more than one photo is used for a specific map.

So the issues with rectifying aerial photos go far beyond compensating for overt censorship. Nevertheless, by exploring the varying use of overtly censored images, it is possible to engage with the practical challenges of the use of aerial photographs for mapping more broadly. Aerial images are censored for Israeli and Palestinian organizations alike, and they hide what are often well-known installations. Yet this censorship has a

Plates I and II

Two maps of Mishor Adumim industrial park. On each map, the large black circle indicates the focus area of Mishor Adumim. Plate I is a detail of an ARIJ (2013b) map of settlement expansion. In the large circle, two military bases are clearly indicated in red, and Mishor Adumim is labeled as an urban area in blue. Plate II is a detail of a Peace Now (2012) settlement map. In the large circled area, no military base is shown, but Mishor Adumim is indicated in blue. In addition, the roads and infrastructure shown in figure 4.3 (and that appear in red in plate I) are even erased from the background of plate II. The size and details of the locations of boundaries for individual areas also vary considerably. Both figures used with permission. Large circle annotations added by the author. The small black circle annotation in plate I appears on the original map and is unrelated to the present discussion.

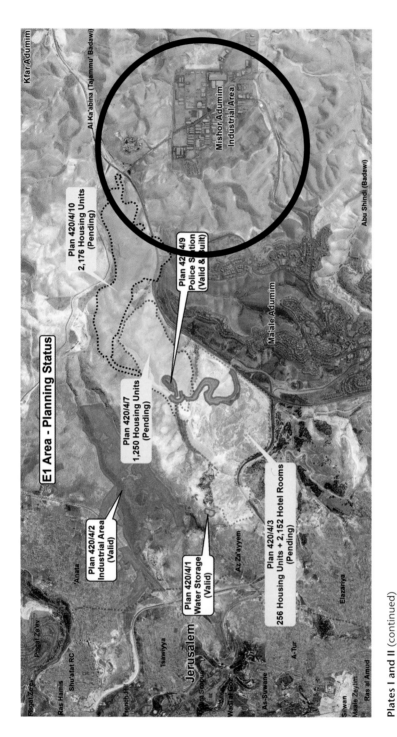

Plates I and II (continued)

Plates III and IV
Plate III is a detail of an ARIJ (2008) LULC map. Crops are shown in shades of green; orange is used for arable land; yellow indicates urbanized Palestinian areas; and blue is for Israeli settlements. The light-tan areas are regions with little vegetation. Plate IV is a detail of a Peace Now (2011) map of Israeli settlements in the West Bank. The map combines data from the PCBS and Israeli CBS in ways that reproduce borders on the ground. The PCBS data show the Palestinian population and appear in brown. The CBS data show the Israeli settlement population and appear in blue. Both figures used with permission.

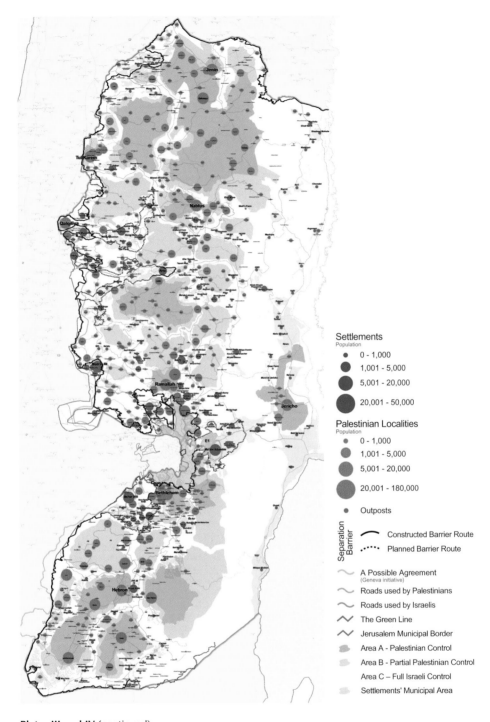

Settlements
Population
- 0 - 1,000
- 1,001 - 5,000
- 5,001 - 20,000
- 20,001 - 50,000

Palestinian Localities
Population
- 0 - 1,000
- 1,001 - 5,000
- 5,001 - 20,000
- 20,001 - 180,000

- Outposts

Separation Barrier
- —— Constructed Barrier Route
- ······ Planned Barrier Route

- ～ A Possible Agreement
  (Geneva initiative)
- ～ Roads used by Palestinians
- ～ Roads used by Israelis
- ∿ The Green Line
- ∿ Jerusalem Municipal Border
- Area A - Palestinian Control
- Area B - Partial Palestinian Control
- Area C – Full Israeli Control
- Settlements' Municipal Area

Plates III and IV (continued)

Figure 5.3
A GIS specialist and civil engineer at ARIJ in the process of rectifying aerial photographs. Photo by the author.

peculiar effect owing to the segregation imposed by the occupation: Palestinian NGO maps frequently indicate Israeli military bases, while many Israeli NGO maps do not. Like the commercial providers of aerial photos, Israeli NGOs must conform to Israeli civilian law, and for this reason they often continue to use the censored aerial images (e.g., figure 5.5b). Certainly not all Israeli NGOs omit reference to Israeli military bases (e.g., B'Tselem and Weizman 2002). But there is pressure to do so, since leaving Israeli military bases on a map would be viewed skeptically within Israel. This poses a particular challenge to Peace Now, as one of its main goals is to change the perceptions of the Israeli public.

By contrast, under the occupation, Palestinian NGOs have little to lose in terms of their relations with the Israeli authorities. For while their use of uncensored data might be met with a harsher response by Israeli authorities, to some extent they live under ongoing harsh treatment anyway due to the occupation. As such, while they might partly depend on

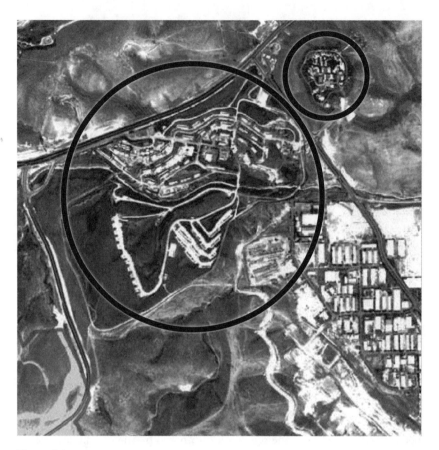

Figure 5.4
An aerial image of Mishor Adumim that shows buildings and roads at the reported site of an Israeli military base. Mishor Adumim is at the right-hand edge of the photo. The large circle clearly indicates buildings and related infrastructure in a region that on ARIJ maps, is marked as a military area. The corresponding spot is shown as an empty plain on Peace Now maps (e.g., figure 5.5b at the bottom). The smaller circle indicates an area that is included in the (civilian) built-up area of Mishor Adumim on ARIJ maps (e.g., figure 5.5a), and excluded from Mishor Adumim on Peace Now maps (e.g., figure 5.5a). This image is based on an excerpt from a UNOCHA (2007, 27) map of Ma'ale Adumim. The European Union provided the uncensored background photograph to the United Nations (see figures 5.5a and 5.5b). Interestingly, as of 2017, these built-up areas are also clearly visible in the satellite layer in Google Maps, which generally uses Israeli images. This map is the responsibility of the author. For updated UN maps of Palestine and Israel, see http://www.ochaopt.org/mapstopic.aspx?id=20&page=1 (accessed April 2, 2016). Circle annotations added by the author.

purchasing images from private companies, Palestinian NGOs are able to combine the Israeli images with lower-resolution information from public international sources, such as the United Nations (e.g., figure 5.4).[32] They do so in an attempt to accurately depict the locations of the numerous military bases and closed zones that encircle ever-greater areas of the West Bank.

The alternating hide and reveal of the military bases also affects the depiction of the locations and boundaries of the settlements themselves. This can be seen by comparing the detail of Mishor Adumim in ARIJ's map in figure 5.5a (top) with Peace Now's map in figure 5.5b (bottom). In Peace Now's map, the region near the top of the large black circled area is not indicated as being part of the Israeli industrial complex of Mishor Adumim, in blue, which lies just east of the settlement of Ma'ale Adumim. In contrast, the ARIJ map of the same area shows a military base in red, and gives a different size and extent for Mishor Adumim, in blue, both located within the corresponding large circled area. There are several reasons why this could be the case, all of which relate to the corresponding within/without positions of Peace Now and ARIJ with respect to the Israeli settlements. It illustrates as well why their positions are not simply an equal mirror, produced in opposition. Instead, this within/without dichotomy serves to systematically privilege Israeli and select international cartographers at the expense of Palestinians, who themselves may be local and/or international.To see how this happens, it is necessary to further explore the details of the mapmaking process for Mishor Adumim in each organization. In Peace Now's case, during its fieldwork, its cartographers would be able to approach the circled region or ask a local resident, and discover that it in fact forms part of the adjacent military base. Then, because Peace Now does not generally show Israeli military bases on its maps, it would have excluded them from the boundary of Mishor Adumim. Indeed, even if the Peace Now researchers were simply unable to enter the area, this too would suggest that it belongs to the base and not the industrial park.

By contrast, ARIJ cartographers' movements were under much greater restriction, and the mapmaking process represented ever-greater dangers. They would not necessarily be able to approach Mishor Adumim to verify whether the area belonged to Mishor Adumim or the nearby base. Thus, they likely would adopt a cautious stance, choosing to indicate the circled

Figures 5.5a and 5.5b
See color plates I and II.

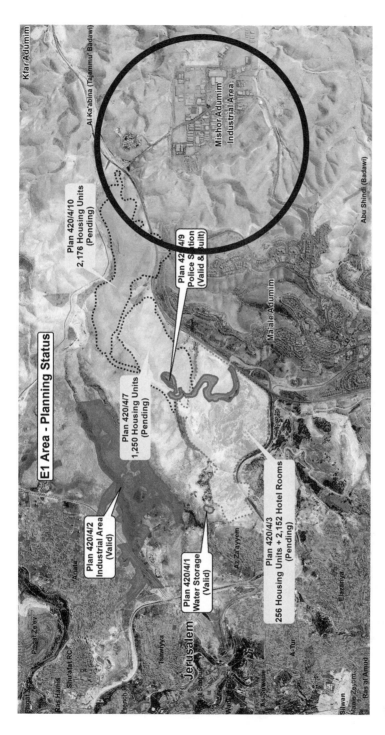

Figures 5.5a and 5.5b (continued)

area as being part of Mishor Adumim versus the military base. So within/ without is far from "separate but equal," to reuse a particularly atrocious phrase from the segregated schools under the Jim Crow regime in the United States. Instead, within/without is a form of cleavage that has as its twin goals the recognition of one party, and elevation of it to access and control over knowledge, combined with the simultaneous subjugation and erasure of another. For in Peace Now's case, its omission likely comes from being granted access—that is, to making a concrete observation of the area's status—while in ARIJ's case it stems from the ambiguities that result from being prevented from doing its work.

However, the shifting boundaries of Mishor Adumim are not just a question of having more or less data in more or less detail. The changes instead extend to boundary determinations at finer scales. Evidence of this can be seen by looking closely at small changes in the boundary of Mishor Adumim in figures 5.5a and 5.5b (see also plates I and II). For every centimeter of boundary, cartographers must make determinations for whether an individual tree, road, field, or pile of stones belongs to the settlement or base, or neither. Their judgment as to whether a settlement or base was actually present would of course affect such determinations. The location of the boundaries is also intricately related to those organizations' geographic positioning with respect to the Israeli settlements, which would affect how closely they could approach the border from the outside, and whether they could approach it from the inside at all.

In this case, although Peace Now's ability to work within the settlements affords it the opportunity for detailed fieldwork, nonetheless that same privilege prevents it from showing the bases or indicating the status of the circled area anywhere on its maps. This thereby illustrates both the complexity and power imbalances of the within/without dynamic with respect to the settlements along with its effects on the empirical content of settlement maps. In the next section, I explore the related but varying implications of ARIJ's position within and Peace Now's position without Palestinian communities.

Mapping Palestinian Communities: Inside Out in Palestine

"We're going back to that Old Testament story, of Samson and Delilah. You know the story?" (interview 11, Palestinian NGO cartographer). As I sat down to analyze settlement mapmaking practices, I kept returning to this exchange that I participated in with a cartographer at ARIJ. For those not familiar with it, Samson and Delilah is the biblical story where an exceptionally strong man, Samson, is betrayed by his lover, Delilah, and chained in his enemies' temple. He then pulls down the roof of the temple with his own chains, destroying the temple and murdering his enemies, but also killing himself in the process. Of course, the use of a biblical story to illustrate contemporary politics is nothing new in many places, including North America. And the adaptation of this tale seemed apt, as a claim that the conflict threatens to destroy the very territory that is one major goal of the fighting. This is not what drew me to it, though. Instead, the very self-evident quality of the story pulled me back. The storyteller didn't claim that the story is identical to the present day, yet he did assert, referring to the relationship between violent conflict and self-annihilation, that "today, it's *practically* the same thing" (emphasis added).

This claim that the past is emphatically not identical but nevertheless commensurable with the present was mentioned numerous times during my fieldwork, and caused me to wonder whether the present had been deliberately made to fit a past, imagined or otherwise. It also drew to mind several more geographical questions: To what degree had the landscape also been made to be commensurable or incommensurable with places elsewhere, and in earlier time periods? To what extent is the process of the production of space, which simultaneously incorporates social imagination and material reconfiguration, itself a self-fulfilling prophecy of the sort described in chapter 4, whereby material and discursive goals coalesce step by step?

The retelling of Samson and Delilah makes a call to the shared international histories that have shaped the lands of the West Bank over the course of thousands of years. Yet the very friction of this coalescence points to another sense of saying that ARIJ is mapping the settlements without. In addition to the material meaning—being outside the settlement fences—there is also the metaphoric sense—going without the types of infrastructure, institutions, ability to avoid interference, and other forms of social and economic privilege that are denied ARIJ under the Israeli occupation.

Although it is relatively quiet at times, the occupation never retreats far from view. The settlement of Har Homa looms over the Bethlehem skyline.

Road closures and checkpoints interfere with its work with more or less regularity depending on the current political situation and obscured military decisions. This sense of "doing without" goes hand in hand with ARIJ's position viewing the settlements from without: looking on the settlements from outside the fence, across the segregated bypass road, and not being able to enter established settlements and outposts at all in most cases, and not casually in any case.

This doing without, however, is related to ARIJ's position within other areas of the West Bank, within the land and its shared stories. The ARIJ cartographers' very exclusion from the settlements—and for many of them, Israel as well—is due to and reinforces the cartographers' positions within the Palestinian communities of the West Bank. This informs its immediate experiences of the effects of the expansion of the Israeli settlement project. Calls come in every day reporting another attack on a farmer, the uprooting of trees, or the impending demolition of blocks of homes. Without documentation, once a particular feature is demolished, it leaves open the possibility that the military could argue that it was never there. In this light, mapping is a way to recover traces that are being expunged, partly in an attempt to recover biblical geographies over and above the many changes that have taken place since ancient times.

One way that contemporary landscapes are being documented is through LULC maps (figure 5.6a).[33] Yet those maps are affected by ARIJ cartographers' position as observers viewing the settlements from without and participating in the Palestinian West Bank from within. ARIJ maps rely both on the cartographers' own experiences on the ground and a coordinated community of local observers. By contrast, the more privileged Israeli cartographers largely observe Palestinian areas from without. There are exceptions to this, such as in East Jerusalem (figure 5.2a), where Ofran sometimes takes pictures of Israeli roadblocks and other obstacles set up to obstruct movement in Palestinian neighborhoods. Those exceptions illustrate the power asymmetries of this constructed within/without dichotomy, highlighting the complexity even within attempts to unilaterally bifurcate space. In what follows, I will explore how these issues relate to the mapping of Palestinian areas. I turn first to how cartographic judgments are grounded in personal and group experiences in the Palestinian West Bank, and second to how ARIJ's ability to collect consistent data goes hand in hand with the development of a network of observers who, like those at ARIJ, have learned to mobilize and build on their collective position of being forced to go without.

Palestinian Communities from within: Visualizing a Politics of Cultivation

As the name implies, land use maps indicate how people are using the terrain—including, for example, by maintaining agricultural fields and urban residential areas—as well as the types of features that cover the land—such as forests, bodies of water, and streets. LULC maps are generally made for planning and management. For instance, a county developer might analyze an LULC map to determine the percentage of low-income housing that lies within a short distance to a river and thus is at risk of flooding. Yet although a distinction is drawn between land *use*, or the ways in which resources are managed and to what purpose, and land *cover*, or what physically is covering the ground, in practice they are often intertwined (Fisher, Comber, and Wadsworth 2005). At times cartographers also have intentionally conflated land use and land cover, in part to disempower those living in the areas that are being mapped. An area of olive trees, say, is both being used for agriculture and covered in rows of trees. So it could be marked as "agriculture" to emphasize the human presence, or alternately, as "forest" to suggest that humans were not actively farming it.

Because they overlap in inconsistent and ethically flexible ways, land use and land cover are frequently mapped together. This does not mean they are identical, however, given that use does not determine cover, or vice versa. An area of rocky outcrops covered in shrubs, for example, might be abandoned—or it alternately could be used for foraging essential wild plants. The two different types of use (abandonment or foraging) nonetheless result in the same apparent cover (rocks and shrubs), and the difference couldn't be seen from an aerial photo. Instead, it would require knowledge of local geographies and agricultural practices.

ARIJ's LULC maps (figure 5.6a) add a level of detail to areas that remained blank on maps for years (see chapter 4). Furthermore, far from being secondary to quantitative facts, land use maps are the very foundation of many statistics. As such, if there is a detailed account of what is present at ground level, then there is a direct line of argument for indicating how much has been destroyed. Mathematical figures such as a tally of total crops burned or determination of what percentage of an area has recently been

Figures 5.6a and 5.6b
See color plates III and IV.

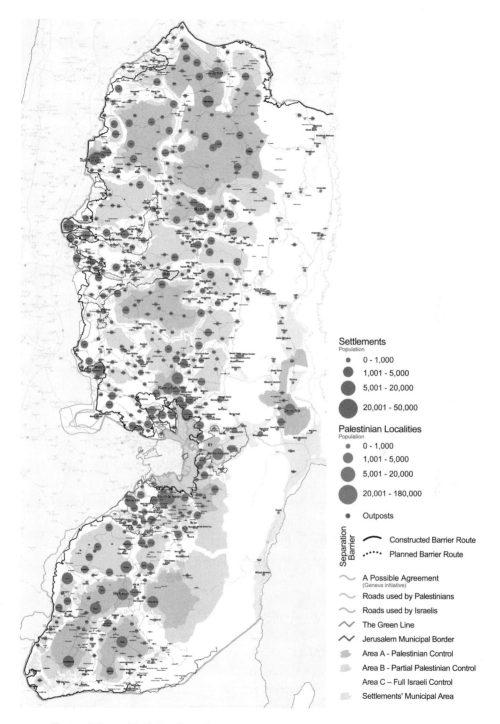

Settitlements
Population

- 0 - 1,000
- 1,001 - 5,000
- 5,001 - 20,000
- 20,001 - 50,000

Palestinian Localities
Population

- 0 - 1,000
- 1,001 - 5,000
- 5,001 - 20,000
- 20,001 - 180,000

- Outposts

Separation Barrier

⌒ Constructed Barrier Route

⋯⋯ Planned Barrier Route

⌇ A Possible Agreement
(Geneva initiative)

⌇ Roads used by Palestinians

⌇ Roads used by Israelis

⋀ The Green Line

⋀ Jerusalem Municipal Border

▦ Area A - Palestinian Control

▦ Area B - Partial Palestinian Control

Area C – Full Israeli Control

▦ Settlements' Municipal Area

Figures 5.6a and 5.6b (continued)

taken over by settlements would both likely come from land use maps—either directly or through a comparison of two such maps over time. Similarly, to find the total area that is covered by settlements, it is necessary to draw borders around individual settlements and find the area within these borders, and then add them to find the area covered by settlements in the West Bank on the whole. These types of calculations are standard in many GIS software packages. The practical impacts of LULC maps are legion given that one of the most ready justifications for the confiscation of West Bank territory by the Israeli authorities is that the land itself is not in use (Shalev 2012).

Land use maps can be made in a variety of ways. While automated processes are being developed to turn aerial photos into LULC maps, and some of these were in use at ARIJ, presently the most detailed and consistent way to make a land use map is to do it by hand on a computer. This involves tracing an aerial photo, or another set of satellite or remotely sensed data, with a mouse. The process is similar to the digitizing of historical maps that is described in chapter 4. It involves opening an aerial photo on the computer, then tracing over the image by hand with the mouse, piece by piece—including buildings, areas of crops, and other noticeable features—thereby incorporating key aspects of the photograph (see chapter 4). As with digitizing maps, it is painstaking and arduous work, but unlike with maps, the goal is not to faithfully reproduce all or most of the aspects in an original map. Instead, it is to "convert" the original satellite photograph into predetermined categories with names like "forest" and "permanent crops." So the outline of a city might be traced, and the whole figure marked "built-up area." As much as it involves empirical observation and the tracing of features, the process of making LULC maps is therefore also intrinsically interpretive.

Palestinian Communities from within: Judging the Land, from Inside an Office

The process of producing the LULC maps required adapting internationally standardized categories to the regional landscape. This is a common issue with using any similar classification scheme, but in this case it relied heavily on the cartographers' combined firsthand experience of the rigid yet shifting geographies of the Israeli occupation. To produce the LULC maps at ARIJ, each cartographer was given a particular district to complete. Once the districts had been assigned, the GIS group sat in one room at individual computer consoles and began tracing photos with the mouse for seven to eight hours per day. It was decided to update the maps by hand precisely because the most recent maps needed to be consistent with the previous years' versions, and those earlier maps had been made before reliable automated methods had become available (interview 9, Palestinian NGO cartographer). The announcement that a team of four to six cartographers would spend months painstakingly making LULC maps using updated aerial photos was greeted with sarcasm and gentle laughter: "Why do the Palestinians always have to suffer the most?"

For similar reasons of consistency, the Corine land cover classification continued to be used. Originally developed to standardize LULC maps in the European Union and beyond, it had initially been chosen to make the ARIJ maps commensurable with European ones (Büttner et al. 2002; Bossard, Feranec, and Otahel 2000). The choice to use the Corine classification, however, involved a further dimension due to the need to adapt and add levels of detail to the categories of the European classification for the West Bank. The application of the Corine classification was by no means a straightforward process, and for this reason, it could not be a solitary one. No categories existed, for instance, to differentiate between Palestinian and Israeli urban areas. These were classified together under the standard category of "urban fabric." Yet subcategories also were created to differentiate them as "Israeli settlements" and "Palestinian built-up areas," respectively. The Wall received a new category all its own, classified (consistent with ARIJ's terminology overall) as the "Segregation Wall," with the subcategory "Wall zone."

However, it was not always clear where the boundaries of a specific feature might lie. Also, in some cases it was evident that the aerial photos, which had to be purchased from Israeli firms due to the prohibition on Palestinians flying planes, had been censored in the manner noted earlier (interview 9, Palestinian NGO cartographer; interview 1, Palestinian cartographer and NGO director). At other times, individual areas of a photo might shift due to the technicalities of the angle at which the photo was taken, and how it was projected from an essentially three-dimensional data set (an image of an arc of Earth's surface) onto a two-dimensional screen. Often, because of intensive land use, two categories were relevant within a single area, which might include, say, vegetable fields (arable land) interspersed with olive groves (permanent crops).

To adapt the Corine classification so that its use was both commensurable with international standards and relevant to the West Bank, the ARIJ cartographers drew on their collective experiences within the physical landscape. The process was often an intersubjective one—an instance of trained judgment (Daston and Galison 2007) to be sure, albeit one where both the training and eventual judgment were decided by coming to a consensus as to how particular elements of satellite images corresponded to what the cartographers had observed during their daily lives. This process of discussing and validating each other's observations was essential, because the maps needed to be comparable across the different districts assigned to each individual cartographer. So the cartographers had to coordinate their decision making with respect to the individual categories. At times, people might work in silence for hours, but usually the office was a lively place, with several people standing around one computer, admonishing each other or making suggestions.

This form of materialized reason was collective on several levels. While the director of the GIS unit was the ultimate arbiter, the stress throughout was on the rationale behind any particular decision. One afternoon, the other cartographers and I were called in to a discussion in the hallway in reference to two versions of the same settlement map, taped next to each other on the wall. Which colors were more objective, the brighter colors or the more muted ones? Which map had the clearest message? The decision to choose the more muted and arguably more scientific colors was made again through an extended conversation involving the entire unit. Although the usual power plays and competitions among different professionals were present, in the context of office politics where there might be friendships or rivalries, the explicit emphasis was placed throughout on what constituted the most objective decision.

In addition to being continually adjusted within the group, the map-making was thoroughly based on broader knowledge of the landscape and its forms. In one instance, when I was unsure whether a pile of stones was a quarry or the remains of an old farmstead, a colleague told me that it was obviously a farmstead, because they are everywhere—the implication being that he saw them all the time in the landscape himself. Indeed, on the walk home from the office, another cartographer pointed out several piles on a nearby hill, asking me to compare them to what we saw earlier in the aerial photo. In another situation, on a field trip to the Tulkarem district with students in a master's degree course in water management at Birzeit University, a colleague from ARIJ poked fun at me by pointing out of the van's window and saying, "Look, permanent crops!" She was referring to the rows and rows of olive groves that hugged the contours of the hillside, and that we would later classify as "permanent crops" on the LULC maps. Back at the office, the olive groves themselves were the subject of much discussion because some were clearly planted in rows from above, while others were more of a jumble. Should they all be considered crops, given that after thousands of years of cultivation, they were almost certainly planted by humans? Or should only those in rows, theoretically newer or more carefully tended, be classified as crops?

In the end, they chose the more frugal option: only the trees in rows became "crops," based in part on assumptions, grounded in the experiences in the landscape, that farmers would systematically place trees with enough room to grow and walk among the rows. Such a decision can also be viewed in a broader frame—as an attempt to counteract previous criticisms that Palestinian cartographers used too broad a definition of cultivable areas. In this vein, Brawer (1994, 341) has argued against including "lower grades" of land in the definition of cultivable areas because the lower grades "have only small scattered patches of arable land, or carry few widely dispersed ... fruit trees." Yet this excision of smaller areas is also arguably an elitist decision in both cases: it ignores the fact, noted by Hadawi, that for survival, all but the wealthiest Palestinian farmers long had to engage in "the cultivation of small patches of soil between the rocks sometimes by means of a pick-axe, or in terracing still smaller pockets and placing olive tree shoots in them." Indeed, as Hadawi (1957, 17) observes, "many village families were able to subsist, though miserably, on such land which, according to Government standards, was classified as 'non-cultivable.'" So efforts were made to farm every miniscule area using heterogeneous methods rather than attempting to impose homogeneous fields on the landscape.[34] In this context, ARIJ cartographers drew on their position within Palestinian communities to form definitions of crops that met international scientific standards more rigorously, in their view, than previous characterizations (largely from without) that cast the entire region as "empty" or "uncultivated."[35] At the same time, their position had varied implications for their maps.

The West Bank from within: Fields Scattered with Maps

If this was the experience in the office, then the Urbanization Monitoring Department took such practices into the field at the scale of the West Bank and East Jerusalem. In contrast to the Geo-Informatics Department, which produced the land use maps that focused largely on agriculture and Palestinian urban areas, the Urbanization Monitoring Department was charged with monitoring settlement expansion. It input detail into the settlements whose boundaries were noted in the LULC maps. This was achieved largely through a combination of aerial photography, remote sensing, and fieldwork.[36] The fieldwork included sightings with cameras with high-powered lenses and observations conducted outside the settlements' perimeter fences. These provided views of settlement architecture from afar that in addition to the less updated depictions of settlement buildings on the aerial photographs at the cartographers' disposal, were their main methods of observing settlements from without (interview 9, Palestinian NGO cartographer). This was not the primary source of their knowledge of the expanding geographic reach of the settlements, however. That came from their ability to act within the Palestinian communities that were directly displaced by settlement expansion to gather the military orders, or leaflets containing maps, that are routinely distributed on Palestinian fields.

As part of these efforts, ARIJ cartographers regularly met and telphoned a dispersed network of contacts (interview 11, Palestinian NGO cartographer; interview 9, Palestinian NGO cartographer). Often the meetings were for the purpose of collecting the Israeli military orders—orders that announce the confiscation of Palestinian land and are vital to settlement expansion. Indeed, the critique of ARIJ that opened this chapter precisely revolved around a report on these types of orders. These military orders are crucial to the settlement project. When settlement construction efforts take place beyond established fences—that is, generally within areas used daily by Palestinians—the Israeli military is required, both legally and practically, to announce further construction efforts to the Palestinians. Throwing photocopies of the military orders directly into the fields was the most expedient—and not to mention disrespectful—method for them to do so. The leaflet drops were also a method of communicating that conveniently, did not usually require the soldiers to personally face the people whose lands were being confiscated.

Each individual leaflet might not provide much information, but collectively the leaflets are valuable data. They serve as a literal paper trail, although more like a scattering of bread crumbs than a defined chain of bureaucratic files. In an interview, one cartographer explained that the simplest manner to obtain the orders was to personally gather them from the field: "When they have a military order, they throw the maps in the fields, and the people collect these maps and bring them to us." When asked if it was possible to collect all the military orders in this way, he answered that while over fourteen hundred orders have been compiled, this only represented a tiny part of the total number. Among the many hazards that prevented the collection of all the dropped records, he mentioned, "Sometimes it was raining" (interview 11, Palestinian NGO cartographer).

In the end, many of ARIJ's intricately detailed maps actually began with cartographers tramping around in fields not just to collect GPS points but also to collect maps. If the Israeli soldiers toss the orders onto the ground as an attempt to dehumanize those living there, it instead has had a galvanizing effect. In a context where the Israeli military prevents Palestinian organizations from having any access to their records of the orders, the scattered maps have served as a trove of information for those without the ability to easily submit requests for the release of data. By mobilizing its position within Palestinian areas, by coordinating its efforts and involving communities throughout the West Bank, ARIJ (2006a) has been able to produce the most extensive freely available database of military orders in the West Bank of which I am aware.[37] So Peace Now and ARIJ have different access to the military data, and this involves differently adapting the methods of collecting those data. In turn, those methods and access differently shape each organization's maps.

In addition to highlighting the importance of ARIJ's position within Palestinian areas, the practice of collecting leaflets indicates a double function of ARIJ's wider contacts. People regularly call to tell the ARIJ employees that orders have been dropped in the fields, at which point the employees drive to the spot to collect samples. They then proceed by interviewing those present, both verifying their observations and distilling facts from the locals' accounts, in part by taking videos and field photographs. These materials assist them when they return to the office to map the changes, which they do through a combination of observations on the ground and from above, of sightings by one and many. In this way, the Palestinian cartographers coordinate among themselves and with the locals to become trained observers and validate or disprove each other's observations. So rather than a transfer of knowledge solely from internationally trained cartographers to farms and villages, forms of observation and validation develop in tandem between the cartographers and local communities even as they work under the daily injustices of the occupation.

This process of validation allowed for the emergence of an account of events that would be recognized as both Palestinian *and* scientific. The empirical quality of the maps allows them to fit the required form of an "account" that is necessary to compare with related Israeli maps in an international context. In contrast to discriminatory assertions that Palestinians were too emotional, the maps provide a coherent picture of the Israeli military's ongoing confiscation of land that would be readily comprehensible to professionals trained in Europe and North America. Empirical maps are one type of map out of many, but they have considerable power because they readily lend themselves to the kinds of statistics and quantification that are privileged in international governance and negotiations. To give one example, I was told that the military can "play games" with the numbers, claiming in its statistics that it confiscates much less land than that indicated on the distributed leaflets. By digitizing the maps scattered with the numerous military orders, ARIJ cartographers can then calculate the total confiscated area and offer counter figures to the statistical claims of the Israeli state (interview 9, Palestinian NGO cartographer). The figures are both commensurable and comparable internationally, even as the process of compiling them stems from the cartographers' and their community members' experiences as Palestinians in an occupied landscape.

In the end, if the military orders inform LULC maps by providing the latest changes in available land, then the LULC maps inform the use of future military orders. They do so by demonstrating that before confiscations, the area was in regular use. As in the office, monitoring land use and confiscation involves extensive knowledge of local geographies—including how to get to sensitive areas without getting arrested—that is brought into fact by the resulting map. The choice to make LULC maps emphasizing the variety of Palestinian land use, observations that constitute the maps, and contents of the maps themselves are all intrinsically influenced both by the status of ARIJ without the settlements and privileges they receive from the

Israeli state, and within the Palestinian communities directly affected by settlement expansion. The position of ARIJ cartographers in the landscapes are thus inflected throughout their LULC maps and related military orders. In the next section, I examine how Peace Now settlement maps are differently shaped by the very different within/without position of Israeli cartographers in the West Bank.

Palestinian Communities from without: The Borders between Data Sets

ARIJ (n.d.-a) "deplores and condemns the death threats against Hagit Ofran, the Settlement Watch Director … of the Israeli Peace Now movement. ARIJ considers that such threats are against international human rights norms." Six months after I first traveled to Ofran's home in an apartment block in West Jerusalem, she stepped out of her door one morning to find graffiti scrawled along the walls of the stairwell. It included threats like "Peace Now, the end is near!!!" Two months later, there was new graffiti outside her door, including the ominous "Rabin is waiting for you." This referred to prime minister Yitzhak Rabin's assassination in 1995 by a Jewish Israeli who was upset about the signing of the Oslo Accords. As I sat down to write this chapter, still more graffiti had been found at Ofran's home, including "Hagit, you're dead." The Peace Now offices have been subject to similar graffiti, bomb threats have been called in at intervals, and the members of the group have received death threats by e-mail, all allegedly from prosettler Israeli groups.[38]

These threats indicate the unique position of Peace Now in general and Ofran in particular. For while they benefit from their incorporation into Israeli and international landscapes, and are able to move within Israel and internationally with less obstruction than many of those at ARIJ, nonetheless their position within has its own, albeit lesser, vulnerabilities. These include the point that with respect to the Palestinian communities, Peace Now must work largely from without. As such, Peace Now is subject to threats from Jewish groups within Israeli society, but also has fewer opportunities to work with Palestinian colleagues who, under different circumstances, might deliver a measure of support despite ongoing disagreements.

One result of Peace Now's position can be seen in figures 5.6a and 5.6b (see also plates III and IV). The figures show, respectively, ARIJ and Peace Now maps. The maps provide a broader view than those of the military bases analyzed earlier in the chapter—an indication of the innumerable borders that are shaped through the collection of data. They also demonstrate yet another way that the borders of the settlements have created metaphoric borders within the offices of the two organizations. The ARIJ map in figure 5.6a includes a wealth of detail for the Palestinian areas (see also plate III). This is far greater than those same areas in the Peace Now map in figure 5.6b (see also plate IV). In both Palestinian and Israeli areas, these differences extend not just to the level of detail but also to the borders and extent of specific regions.

However, this lack of detail is far from the only effect. The influence extends to the choice and composition of data sets themselves. In cases where Ofran wants to map Palestinian areas, she must use Palestinian data, because her ability to collect data in Palestinian areas is somewhat limited.[39] Yet she cannot always obtain data directly from Palestinian organizations like ARIJ, because it too is located within Palestinian areas of the West Bank, not to mention the climate of suspicion due to the occupation. Thus, she must find sources that are publicly available online, such as that of the PA's PCBS, whose work is described in chapter 4. This reproduces implicit borders within a single map, which can be seen in figure 5.6b, where Peace Now combines Israeli government data with PA data. On the map, the data for the West Bank Palestinian population are taken directly from the PCBS, while the data for Israelis are from the Israeli CBS. In so doing, the map seems to acknowledge Palestinian sovereignty in the collection of knowledge of its own populations, but also reinforces diverging representations of Israel and the West Bank, with varied political effects.

The focus, scale, detail, and composition of data thus mimic the sphere of mobility of the cartographers themselves, as the "holes" in their landscapes are filled by the knowledge of groups that are partly produced through that very segregation. These are also reproduced in the use of aerial photographs and international data such as that from the United Nations that are nonetheless also situated within political divides rather than beyond them. So the boundaries on the ground are reproduced by boundaries in the data. Segregation is recursively reinforced. The very separation of groups, by divorcing their bodies and leading to the development of bifurcated scientific communities, thereby shapes their observations—and subsequently, their data collection and display. This can further differentiate the goals and strengths of the data collected by each group, and that in turn has implications for cross-border work (interview 18, Palestinian NGO researcher).

These challenges don't mean that implicit alignment, a form of productive *noncooperation*, is impossible, though. For example, an ARIJ (2000; see also 2006b) report on the expansion of the settlement block of Modi'in includes a table whose sources are listed in combination as the "ARIJ database" and a "Peace Now Settlement Watch report." While public shows of support can be difficult on an organizational level, it is not entirely unusual for staff members or related academics and activists, as independent citizens, to appear across sides on panels or in debates (e.g., Ofran et al. 2011), on news outlets (e.g., Eviction in East Jerusalem 2009), or to be published in the same journals and edited books (e.g., Ma'oz and Nusseibeh 2000). Like Peace Now's use of PCBS data in figure 5.6b, noncooperation resembles an agreement of noninterference that incorporates a working conception of alterity, and in the spirit of the recasting of triangulation in this chapter, acknowledges the limits of one's own knowledge. It demonstrates both the restrictions and innovations of work under occupation.

Validating the View from the Top of the Wall

If noncooperation can be productive, then it must do so within an atmosphere of deep suspicion. Words that have benign meanings elsewhere can have particularly dismal connotations in the context of occupation. *Normalization*, which might imply a sort of settling down, a return to the rhythms and regularities of life, instead refers to the acceptance of the oppressive status quo, or any act that encourages a view that the occupation is normal, that naturalizes its seeming inevitability.[40] Similarly, the word *peace*, after rounds and rounds of failed negotiations, has connotations of stagnation and false promises. So the mere mention of peace can have a chilling effect. Some are able to experience this form of pyrrhic peace by walking in divided Hebron, for instance, or proceeding along the abandoned roads of shuttered shops that line the Separation Wall.

This is the upside-down world (Galeano 2000) in which Palestinian organizations and Israeli organizations could seek to work together, as asymmetrical segregation reinforces dominating realities and knowledges. ARIJ and Peace Now are differently positioned in the landscapes of the West

Bank, and draw on the unique sets of experiences that result from their unequal positions. Their experiential backgrounds in turn shape the kinds of observations that their cartographers are able to make, with the result that the geographic borders that the cartographers seek to map themselves intrinsically filter into their maps.

For their part, ARIJ cartographers make do with being without in terms of the settlements. Yet in mapping settlement expansion beyond existing fences, the cartographers also have access to their own experiences in the land and a network of observers throughout the territory—both of which are only possible due to ARIJ's recognition as an organization that operates within Palestinian communities.[41] In contrast, Peace Now cartographers benefit from a privileged position in both Israeli and international land-scapes, and the access to the settlements from within that this affords them, although they simultaneously must contend with mapping Palestinian areas from without. This chapter has offered a critique of the multiple, over-lapping, and frequently damaging impacts that the power asymmetries of segregation can have on knowledge of that segregation.

The efforts to maintain such folded and complexly spatial within/ without dichotomies thus demonstrate not only separation but also the systemic hierarchy that affects Palestinians, who routinely face the injus-tices of living under occupation. The power imbalances are at once highly unequal, since they vastly disproportionately hinder Palestinians, and more complex than any one attempt to impose a dichotomy might allow. In this light, undoing the segregated landscapes alone will not be effective unless the privilege within shared spaces is as actively addressed.

In addition, by pointing out the consequences of the segregation endured under the occupation, I do not mean to suggest that free move-ment is the proper alternative. Indeed, the current landscape enables the "free movement" of the settlers, and this type of freedom might just imply the ability of those in power to go where they please while imposing their will on everyone else.[42] Like the term *peace, free movement* has lofty conno-tations, and it is often posed as an ideal that is marred by segregation. Yet although greater mobility can be to the benefit of all, under such asym-metrical material conditions as those of the occupation, mobility would not be truly free even without walls or other barriers. Instead, in the region and beyond, freedom of movement might simply allow the powerful the opportunity for freer domination.

Rather than aiming for freedom, the ARIJ and Peace Now cartographers negotiate the challenges of repositioning themselves in ways that serve to question the notion of *position* itself. Certainly, in the view of many similar observers, the Wall presents an incredible hindrance that should be immediately dismantled. Such a dismantling is not the whole story, however, for conceptions of mobility must also confront questions of power and control among people who comingle without physical barriers. Thus, the reflexive production and use of space can allow for greater attention to its effects on knowledge and related social justice efforts. Although the movements of cartographers and others will always be situated in landscapes, this does not necessarily need to consist primarily of forms of subjugation. I conclude this chapter with three brief examples of how fostering reflexivity in space can provide important correctives for and ways to build on the analysis of settlement maps presented so far. I analyze them by formulating the notion of *refraction*, which commonly refers to the ways that light changes when passing through a medium. In social science, *reflexivity*, which could be seen in relation to both *reflex* and *reflection*, refers to the work of ongoing self-observation and self-awareness.[43] *Refraction*, then, implies similar work to enact an awareness of the material and spatial situatedness of an individual, group, or discipline.

Refracted Maps

The first example explores ways that existing geographies might be reimagined from within through the practice of refraction. These include how community workers have sought to refashion forcibly segregated spaces, changing them from spheres of domination into positively self-segregated areas of community solidarity. In response, some Israeli activists have acknowledged the importance of Palestinian spaces by strategically staying out of them, thereby aiming to rework their own privilege. The second example deals with the work of activists who have sought to differently inhabit the geographies of the occupation by slipping through the rigid separation barriers. This involves a way of moving differently, not solely in opposition to the passages of the segregated landscape, but by finding loopholes in those pathways. The third example relates to the alternative method of triangulation used in this chapter. The analysis that I provide seeks not to operate as a "view from above, from nowhere" (Haraway 1988, 589) that seeks superior ground in order to critique Israelis and Palestinians.

Instead, it is an attempt to engage with my own privilege as an outside international in order to argue for a better understanding of how international knowledge partially contributes to rather than alleviates the extreme conditions of the occupation.

Of benefit to the first example, Palestinian communities at times have been able to turn the challenges of segregation on their head through a form of positive self-segregation. In the process, they refashion the meaning and use of existing territories. So if positive terms like *freedom* can imply oppression, negative terms like *segregation* also can be turned upside down. As in the community networks that ARIJ mobilizes, producing networks and spaces as distinctly *Palestinian* ones can further group solidarity and community building. This refractive use of space is especially marked among some groups of Palestinian refugees. Scholars have pointed to how in particular camps, refugees internally organize themselves according to village or region. This is partly a means of coping by reproducing familiar geographic orders, albeit with varying effects (al-Araj 2008, 25; Bshara 2012; Sayigh 1977; Taraki 2008b). Yet it also indicates an ability to build geographic memory into contemporary places in ways that changes how people organize, produce, and move through space. While the politics and practices of maintaining such spaces can be incredibly difficult (Smith 2013), they demonstrate how attempts to alter positions and mobilities can transform one's own subjectivity in place, and in this way can alter the practice of observation itself.

Therefore, a different type of liberation might be obtained within oppressed groups organizing with some independence and control that is separate from the privileged. Elisha Baskin and Donna Nevel explore this notion. These two Jewish antioccupation activists have argued for self-segregated spaces on the part of Jewish activists who are privileged in the context of joint Palestinian–Israeli solidarity work. They aim to avoid turning "Palestine solidarity spaces into support groups for Jewish activists" (Baskin and Nevel 2013). Ofran herself also acknowledges it in an interview with Al Jazeera English. In response to a question about voluntary relocation for Palestinians evicted due to settlement construction, she notes, "That is up to the Palestinians who live there to decide. I cannot decide for them" (Eviction in East Jerusalem 2009). In this light, Ofran's acknowledgment of Palestinian organizations suggests that respect be paid to their self-definition. But the success of self-segregation relies on who has control over their own ability to self-segregate, since actions that in one case can seem like a respect for self-determination are in others what might become a form of abandonment and abnegation of responsibility.[44]

Second, the refractive use of landscape is present in antioccupation activism when protesters find ways around imposed segregation. In the process, they form relationships and build up support networks in Palestinians towns like Nabi Saleh and Bil'in.[45] The Wall and entire apparatus of the occupation have daily damaging effects, and have been thoroughly criticized for the severe impact they have on daily life. Yet the segregation is not absolute, for there are numerous areas that fall within or in between borders. These include slips of space between particular settlements and Palestinian towns where activists sometimes are able to meet discreetly as well as temporary activist spaces. One of these was the Palestinian settlement of Bab el-Shams, which was set up on private Palestinian land to draw attention to the double standard of settlement policy—and was subsequently demolished within forty-eight hours by Israeli authorities (Nasser 2013). Moreover, out-of-the-way sections of the wall are regularly climbed—albeit in considerable physical danger—for a fee by means of a ladder and rope. This crossing over at least has the potential to reconfigure communities and a sense of belonging beyond a dichotomous frame.

Yet even the danger of accessing such areas, due in part to visa regimes, and ability to make productive use of communications technologies differs based on cartographers' social and material positioning within the international geographies. This connects to the third way for refractively engaging with landscape: namely, through the alternative method of triangulation proposed in this chapter. In contrast to its use in surveying, where triangulation was employed to produce a universal view or network, I have used triangulation as a method for acknowledging the ways that this analysis of settlement cartography takes place relationally within the geography of occupation in ways that might transform the observer, the observed, and the act of observation itself.

Triangulation is an attempt to grapple with a feature of internationalism noted in this chapter as well as the chapters that preceded it. Specifically, the first impulse of international scientists under conflict conditions is sometimes to denigrate those from the region—those who often have the most experience working in the specific contexts under study. Thus we are told that Palestinian scientists are "doing it wrong" or Israelis are "manipulating" figures for their own ends.

Certainly the use of science purely for its legitimating powers is present, as it would be almost anywhere. However, the method of triangulation, of analyzing practices across groups, emphasizes the effects of the landscape on attempts to translate facts and figures from one position to another as well as on producing those positions. It draws scholarly attention to the power dynamics implicit in the act of observation among different groups that are facing questions of their own legitimacy in their practice and lives more generally. It requires refractivity in the analysis of those observations—an attentiveness to all the relevant observers' broader positionings. Given this, it highlights the incredible amount of work that is necessary before Palestinian census data can be put to use in the Peace Now offices in West Jerusalem or the Corine classification can be refashioned for its use in ARIJ's LULC maps. Such refashioning of standards is common. But in this case, it includes additional informal local knowledge of the geographies of the occupation together with the necessary professional expertise, required community connections, and ongoing challenges faced when Palestinian cartographers collect data on Israeli settlements on the ground.

Likewise, when I say that I triangulate my analysis with Palestinian and Israeli cartographers, I seek to acknowledge that critical analysis, like empirical analysis, is shaped in landscapes. International academics—at least those without long-standing personal connections to the region—do not have access to any omniscient overview of the region and its struggles. Instead, we both impose ourselves and are drawn in. This requires tremendous effort on the part of organizations that strive for their maps to fit the empirical form that would be readily intelligible to those trained in Europe and North America. They work not only to have a place or voice in debates and negotiations but also so that their efforts might be recognized as a voice, and that the lands they inhabit might be acknowledged as viable places at all. Indeed, the very role that scientific knowledge plays in this analysis—a standard to be conformed to, and a specter of an international audience—is evidence of this power.

Yet the very privilege of those internationals who can travel across multiple borders can mask the ways that landscapes shape the mobility of privileged. Even if they were aware of the difficulties faced by Palestinians seeking to cross into Israeli-controlled areas, a hired consultant who travels in a single day from Oslo to Ramallah (for example) would likely develop a different embodied knowledge of travel under the occupation than would someone who did not have that privilege of movement. Indeed, some areas and aspects of the occupation are made invisible to privileged internationals. This has the effect of reinforcing the continued naturalization and effacement of hierarchies of knowledge. It allows those with privilege to proceed, problematically, as if international science and technology were at once placeless and applicable everywhere. One way to incorporate an awareness of this and related issues into research is to allow greater space for alternative practices, and pay concerted attention to variation among scientific communities in the spirit of the alternative form of triangulation that I have proposed. Overall, the practice of refraction, combined with critical triangulation, serves to demonstrate how intricately and intrinsically geographies of occupation are woven back into knowledge of the occupation itself.

6 The Geographic Production of Knowledge

[To a semi-Orientalist:] If it's what you think

supposing now that I am stupid, stupid, stupid

and that I don't play golf,

don't understand high technology,

and cannot fly a plane!

Is that why you took my life, to make

from it your life?

— Mahmoud Darwish, حالة حصار [State of Siege]

[إلى شبهِ مستشرقٍ:] ليكُنْ ما تَظُنُّ

لنَفْتَرِضِ الآن أني غبيٌّ، غبيٌّ، غبيٌّ

ولا ألعبُ الجولف،

لا أفهمُ التكنولوجيا،

ولا أستطيعُ قيادةَ طيّارةٍ!

ألهذا أخَذْتَ حياتي لتصنَعَ منها حياتَكَ؟

In the above excerpt from his book-length poem *Halat Hisar* [State of Siege], Mahmoud Darwish (2002, 73) references the connection between Orientalism, a regime built on geographic and cultural discrimination, and a belief that as a Palestinian, he might lack technological skills.[1] He poses the question: is a lack of an ability, different method for using technology, or preference for things other than technology worth taking his life? If so, then it implies that his life is worth less than the lives of those who know how to use technology. But in that case, why use it as the foundation for another's life? If what he has (his life) is allegedly so worthless, because he's supposedly not "advanced" enough to use technology, then why bother taking it at all?

The poem critiques "semi-Orientalists" for, on the one hand, devaluing the poet while, on the other hand, seizing his land and culture, and using it as the foundation of their own. This excerpt is representative of Darwish's work more broadly,, which is replete with references to technology, maps, and borders. Two of his most well-known poems are titled, respectively, "Passport" and "Identity Card."[2] In another, lamenting his estrangement from the reshaped landscape of Palestine and Israel after 1948, Darwish (2004, as translated in Antoon 2008, 224) refers to a "map of absence" and notes the distance he feels "when I approach the place's topography." Similarly, in the poem "Earth Presses against Us," Darwish (2005, as translated in 2003, 9) asks the rhetorical question, "Where should we go after the last border?"

His poetry evokes feelings of loss for the land that has been destroyed since 1948, for places that have been plowed through and ground under, and whose inhabitants are engaged in an ongoing struggle to exist. Darwish is one of the most famous Palestinian literary figures of the past century, and his work is widely popular in a region where poets generally are as well known and venerated as musicians are elsewhere. He has written on themes that are present in every chapter of this book. One example is Darwish's (1993) critique of spurious claims that Palestinians don't exist, in the poem "I Am There" from the collection *Ara Ma Urid* [I See What I Want]. During his lifetime, he repeatedly engaged with the iconic way that maps were inflected throughout Palestinian struggles, which often took place in and through land that was continually remade to fit cartographic borders. As such, one of his collaborative volumes was appropriately titled *Victims of a Map* (Adonis, Darwish, and al-Qasim 1995; see also Wood 2010).[3]

Darwish's work illustrates how differently the contemporary geographies of Palestine and Israel appear to those who are deeply conscious of the lives and livelihoods that have been suppressed in order to create and maintain those geographies. While this suppression is sometimes presented as only affecting Palestinians, I have shown in this book how it differently influences all knowledge, including international technoscience, by feeding back into empirical mapmaking practices in sometimes-unexpected ways. Building on STS conceptions of social construction, I have analyzed the *geographic production* of knowledge in order to demonstrate how the power imbalances *in* landscapes shape empirical maps and borders *of* those landscapes. Palestinian and Israeli societies are often studied separately—not without reason given their cultural, historical, linguistic, and other differences. Even so, there is also merit in researching Palestinian and Israeli cartography together as part of a symmetrical study, of examining the relations between them without asserting that they are either wholly commensurable or wholly incommensurable—of investigating how borders occupy not just land but knowledge as well. Cartography in the occupied Palestinian territories is sometimes significantly different from that in the annexed areas of Israel, but the two are joined through the highly asymmetrical relationships of both the Israeli occupation and international technoscience. They also influence both the occupation and technoscience more widely, illustrating one way that the injustices of the Israeli occupation resonate far beyond Palestine and Israel.

It is well known that segregation creates separate communities and invariably privileges one over the other, to the point that the very existence of the oppressed community may be erased. Less well understood are the ways that unequal geographies can limit the knowledge created in and about those communities. But scientific and technical knowledge cannot help to bring about an end to segregation and occupation if the production of that knowledge is also segregated and asymmetrical. The population, governance, and urban maps analyzed in this book combine to show how the work of empirical researchers, and especially Palestinian ones, is unduly discounted due to the material imbalances of geographic landscapes. This has reinforced the exclusion and estrangement of many, in ways that Darwish so iconically describes.

The previous chapters have laid out several of the means by which the landscapes of occupation serve to shape empirical knowledge. This shaping undervalues Palestinians and their knowledge forms in three particularly notable ways. First, Palestinians' right to produce knowledge, to be acknowledged in terms of their very existence and ability to call themselves Palestinians, is continually challenged. So in chapter 3, it was seen that Palestinians have strived to demonstrate their own existence even as Israeli population maps were changed to take account of it. Second, Palestinian cartographers' practical ability to build knowledge institutions and infrastructures is subverted in ways that continue to affect the content of even the simplest governance maps. Thus, chapter 4 dealt with the challenges facing Palestinian institutions, from raids to restrictions on data. Third, as argued in chapter 5, even when Palestinian organizations are relatively well funded and frame their assertions to fit dominant conceptions of knowledge, the content of that knowledge is sometimes considered unverifiable and therefore invalid because it reflects the distinctive experiences of Palestinians *as Palestinians* in the field. So Palestinian professionals are less likely to be accepted as credible observers even when they work fully within the paradigms and methods of international science. Understandings of geography and maps among Palestinian writers, artists, and scientists—including literary figures like Darwish—in this way illustrate the fundamental challenges faced by those who seek to refashion landscapes of oppression, allowing for a more textured understanding of the social shaping of scientific and technical practice.

This does not mean, however, that Israeli cartographers are all powerful. As chapter 5 also shows, the occupation differently shapes Israeli (and for that matter, international) maps. Indeed, in a context of broader anti-Semitism in the 1950s and 1960s, chapter 3 discussed how Bachi aspired to meet the requirements of international science. Chapter 4 touched on the persistent British colonial legacy that also resonates throughout Israeli maps, and chapter 5 demonstrated how Peace Now cartographers were forced, in no small part because of the occupation, to work outside Palestinian communities. These issues pale in comparison to the challenges faced by many Palestinian professionals, but still the consequences for knowledge are both forceful and obdurate.

Together, these chapters show the debilitating effects of the Israeli occupation on empirical knowledge writ large. Yet they also indicate that Palestine and Israel are far from an exception to the international practice of science. Instead, they are a core case of the differential shaping of empirical technoscientific practice around the globe. International scientific and technical knowledge is always situated in landscapes. As such, the Israeli occupation also influences the efforts of international academics and NGO workers, including critical scholars. The cases here therefore serve as a reminder against assuming that any knowledge might exist independently of how it is situated—a caution against facile characterizations of people as practitioners of "bad" science, without attempting a more thorough conception of the landscapes where they live and work. A deeper investigation of how knowledge production is simultaneously material and social thus has fundamental import for the very act of understanding, itself.

The Significance of Negative Space: Remaking the Master's Tools

Darwish's words recount the losses he experienced after 1948 and 1967. In so doing, they serve as a reminder of what no longer is and what might have been. They bring imaginations of lands and people into being for those with no experiences of these landscapes prior to those dates. They also point to the role of technology in their dispossession, and as such provide an interesting counterpoint to the words of a very different thinker—the poet and writer Audre Lorde. Darwish was a figurehead of the Palestinian national struggle, and Lorde was a Black lesbian feminist in the United States. Lorde dealt not with the ways that technology erases people but rather how domination itself could be built into the fabric of technology. But together, their alternative understandings of the force of science and technology are more relevant than has yet been acknowledged. Indeed, both were attentive to the fact that technoscience includes forms of dominance so intricate that they are even adept at hiding themselves from privileged critics of technoscience.

In the remainder of this concluding chapter, I will consider some critical implications of Darwish's and Lorde's work, together with that of one additional theorist, Egyptian feminist Nawal El Saadawi, in order to demonstrate the broader ramifications of the arguments put forward so far in this book. I examine these perhaps-unexpected authors precisely to illustrate what is lost when researchers too readily dismiss knowledge from the margins of critical thought. Such a dismissal is evidence of a form of social and material elitism that presents oppression as a lack of ability, with the consequence that those outside the material privilege and social networks of academia are judged as being less rigorous thinkers, and their potential contributions go underacknowledged.

In the process of doing so, I also return to the three themes from chapter 1 of *internationalism*, *symmetry*, and *landscape*, albeit here in a slightly different order and specific form: *international knowledge* (internationalism), *background* (symmetry), and *material hierarchies* (landscape). Before turning to these themes, though, it is crucial to further contextualize the choice of these three authors. To begin with, this choice is not meant to align the book fully with their points of view. Many researchers have embraced notions of the relationality of sex and gender, helpfully moving away from more essentialist formulations popular in the second-wave feminism of the 1970s and 1980s that provided the context for their work.

Lorde is a pivotal thinker for those who study how different forms of oppression—for example, those based on gender, race, and class—can intersect. She analyzed second-wave feminism in detail, drawing out its lack of attention to the kinds of privilege that fuel racism and classism in particular. El Saadawi contributes a biting analysis of injustices faced by women in the global South—work that unlike her more widely known criticisms of gender and sexual politics in Egypt, has received less concerted attention in Europe and North America. Nevertheless, by positioning herself as a secular writer, El Saadawi unfortunately has also sometimes downplayed important insights from Islamic feminists and related thinkers in the anthropology of Islam. By drawing on Lorde and El Saadawi while still learning from subsequent critiques of their work, I seek to show several ways the work of broader communities of thinkers can contain much of value for theories of technoscience and society. Thus, for instance, rather than implicitly adopting a taken-for-granted national or geographic scale, this book treats Palestinian, Israeli, and scientific identities and practices as being both fully intersectional as well as differently positioned within international systems of power.

However, it is not necessary to accept someone's perspective wholesale in order to benefit from their insights. I use these writings also to illustrate how alternative conceptions might be productively, if selectively, taken up.[4] Certainly, a similar selectivity is practiced regularly with respect to canonical thinkers such as Karl Marx or Max Weber, who continue to be used in new and interesting ways, even though many would now consider their thoughts on race or gender, though common enough in their respective contexts, to be reprehensible. Yet unlike cartographers who are forcibly confined to cells and small areas of the landscape, researchers in Europe and North America have too often voluntarily restricted ourselves only to using the writings of other scholars working in Europe and North America. Even among postcolonial and critical social theorists, greater deference continues to be afforded to such canonical thinkers as well as financially elite scholars producing canonical forms of knowledge. The argument put forward usually suggests that the canonical thinkers' work was merely "better," but this doesn't always hold up to further scrutiny.

Instead, as I have sought to show, judgments about quality are shaped by researchers' positions in multiple landscapes. They are shaped in ways that require greater attention to knowledge's materiality, such as through the study of how knowledge is geographically produced. For just as cartographers and scientists are differently situated, so are critical academics. Certainly, as with any book, this one is thoroughly imbued with the benefits and challenges of my own specific positionings. As noted, however, this does not mean that I discard mainstream thinkers whose work, it almost goes without saying, has helped to make this research possible. The preceding pages are replete with references to their publications. Rather, it requires a turn toward greater material and spatial reflexivity. As noted in chapter 5, this involves ongoing *refraction*, as I have termed it, including efforts to foster an embodied and material engagement with the ways that landscapes of oppression and privilege differently as well as relationally shape the production of academic knowledge (Mohanty, Russo, and Torres 1991). This ranges from the sources and scholars cited, to the sometimes-dismissive characterizations of those sources out of context, to the lack of awareness of geographically diverse histories and institutional patterns and practices of critique.

This shaping need not be intentional or even self-aware. Many researchers care deeply about the societal relevance of their work, whether immediate or ultimate. Yet privilege and oppression combine in ways that are essential considerations for anyone concerned with varying forms of knowledge and the systems that help produce them. The ethnography informing the preceding chapters was aimed at taking seriously the efforts of cartographers in the region, therefore giving them similar space and consideration to that afforded to academic theorists. This chapter is a call to further extend the goal of achieving reflexivity and refractivity in scholarly work, yet with an awareness that doing so is as theoretically challenging, and prone to potential failure, as the most intricate mathematics and densest analytic philosophy.

Asymmetrical Backgrounds

The master's tools will never dismantle the master's house. They may allow us temporarily to beat him at his own game, but they will never enable us to bring about genuine change.

—Audre Lorde, *Sister Outsider*

Lorde (2007) made this remark to the Second Sex conference in New York City in 1979, and it was later reprinted in her groundbreaking collection of essays, *Sister Outsider*. Her conception is in alignment with the later publication of Langdon Winner's (1980) separate formulation that is well known in STS: "Do Artifacts Have Politics?" In her piece, Lorde alludes to the history of slavery in the Americas, arguing that it is necessary to develop different kinds of tools to end injustice instead of simply replacing old masters with new ones. Indeed, in chapter 2, I explore how maps and cartography are in many ways iconic tools of domination, for they enabled widespread forms of imperialism and colonialism that continue, albeit in relatively different guises, up to the present day.

Since Lorde first spoke, her notion of the *master's tools* has been used widely in critical theory. In diverse texts, it is taken to mean any method that comes from and supports existing systems of power, particularly by perpetuating dominant forms of discourse and knowledge. In light of this general conception, GIS mapmaking can be considered a master's tool. Yet Lorde herself used the phrase in a more restricted sense, referring to the paltry efforts to reach out to Black academics among the organizers of the conference at which she was speaking. This can be seen as an effect both of personal agency or individual decisions as well as how systemic racism contributes to a form of academia that consists predominantly of white researchers who only know other white researchers. So Lorde intended the master's tools to refer not only to concepts or practices but also to how the movement, networks, and concentration of resources among academics themselves can affect the constitution of knowledge.[5]

Lorde's more specific notion of the master's tools allows for a fuller exploration of how technoscientific practices that do not address material imbalances, of the types examined in this book, often serve to reinforce the exclusionary aspects of dominant forms of knowledge. Together with Darwish's highlighting of how technological ability is framed as individual inadequacy, rather than in light of systemic economic and social hierarchies, this insight addresses the theme of internationalism that was discussed in chapter 1. In addition to its relevance to Lorde's specific notion, internationalism's applicability to the more general conception of the master's tools is perhaps readily apparent. Throughout the book, I have analyzed the ways that in Palestine and Israel, cartographic practices have served as both empirical tools and instruments of international and imperial domination. I have argued for greater attention to how international landscapes shape knowledge about Palestine and Israel as well as for a more in-depth account of the role that landscapes play in the production of international technoscience. Furthermore, I have studied how cartographers in government agencies and related organizations attempt to use cartographic methods, including GIS, to stretch the boundaries—among them, the geographic boundaries—of technologies and the epistemologies they embody. For this reason, one implicit theme throughout has been that landscapes might be remade in order to facilitate the refashioning of the master's tools into implements that might be used more effectively in current struggles for social justice.

Doing so requires a more thorough understanding of the varying roles of international knowledge. While the international influence in Palestine and Israel is a continual subject of the history and political theory of the region, the role of internationals is frequently sublimated in studies of science and technology in the conflict. Forms of international control are often implicit in standards that govern the supposedly right way to practice science. Thus they continue to exert significant authority in determining how areas are known, with the result that local knowledges continue to be downplayed and discounted. Yet scientific and technological practice is neither ever fully localized nor fully global. This requires further study of the erasures in narrowly liberal forms of internationalism through research on the geographic production of knowledge that is attuned to spaces that are both relational and discrete.

Additionally, this points to the need for international actors and organizations to deal more concertedly with the legitimating power of technoscience, as is evident across the case studies in this book. Bachi often looked to the international scientific networks of which he was a part in order to legitimize his statistical claims (chapter 3). However, he was also constrained in terms of how much he could vary from idealized notions of science that were sometimes ill suited for examining the geographies of Palestine and Israel after the onset of the 1967 occupation. Like the settlement mapping organizations in chapter 5, the PA agencies in chapter 4 were consistently subject to the judgments of international bodies in terms of whether or not they were seen as being fit for self-rule. Such judgments have had decisive impacts on political negotiations and the possibilities for the future funding of the PA. Yet historic British colonialism and the ongoing Israeli occupation also shaped Palestinians' access to infrastructures and landscapes in ways that have largely gone underacknowledged by those actors charged with funding and evaluating their work. As a result, responsibility was frequently shifted onto the PA by way of informal accusations that it was not capable or willing to practice "real" science in the "right" (read: dominant international) way. In this context, it becomes clear why Darwish's question would resonate so widely. Is his inability with technology really worth ransoming his life?

Alongside internationalism, Lorde's more specific sense of the master's tools in relation to the practice of material *backgrounding* has further relevance here. In chapters 1 and 2, I examined the theme of symmetry and argued that it was necessary to temporarily place two groups, Palestinian and Israeli cartographers, side by side—not to compare the two, but instead to highlight the material asymmetries between them. In this chapter, I would like to recast symmetry by turning to Lorde's awareness of how symmetrical imbalances, such as those between Palestinians and Israelis, also can lead to the resilient erasure of one group from knowledge itself. I conceive of this material and spatial aspect of erasure as a form of backgrounding. As I use it, backgrounding is a practice that includes but is not limited to obdurate methods of either forcing people into the background or even just allowing them to persist starved, as if they were too insignificant to deserve mention. In this vein, Foucault (1998, 138, also see Selmeczi 2009, 521) famously referred to biopolitical practices as the power to "*foster* life or *disallow* it to the point of death." Landscapes are a prime example of this conceptual backgrounding, as they are frequently taken to be the empty arena for individual movement. Roads and land cover, such as forests or urban areas, are drawn onto empty paper, a blank screen, and the very "blankness" of that screen is critiqued throughout this book. In chapter 3, the choice to not include data about many Palestinians led to their towns being depicted as blank on Israeli population maps, for instance.

In contrast to Lorde's perhaps more deterministic formulation of the master's tools, however, throughout this book I have sought to demonstrate that these effects are both persistent and even pernicious, but entirely subject to change. After all, an artifact or system that tends to support those in power need not necessarily do so under different conditions or with modifications in its design. Yet the efforts to reshape artifacts, like GIS, meet with greater challenges, including economic and geographic ones, than is typically acknowledged.

The relationship with landscape is far from the only form of backgrounding. Indeed, a related form, violence, can be seen as the complement to Willem Schinkel's (2010, 220) observation that "if the social sciences are blind to their own violence, they do violence to themselves." In Schinkel's view, this is so because social scientists sometimes ignore an intrinsic aspect of contemporary research—namely, the violence that is involved in selecting and separating out any object of inquiry. Another violence pertains to the selection and separating out of academic institutions themselves. In Lorde's case, she refers to implicit exclusion through the violence of institutionalized inaction, the ongoing maintenance of the exclusionary constitution of academia through a lack of effort and reflexivity, even among progressive academics. To be fair, those same progressive academics might be frozen into inaction because of an awareness of the problematics of diversifying. Still, rather than allowing inaction to maintain the status quo, such concern over the challenges of reshaping institutions deserves more attention in academic inquiry. This is seen, for example, in calls to further examine the close ties between many universities and the Israeli military—ties that are discussed below.

In Palestine and Israel, backgrounding is effected in part through the enforcement and maintenance of material hierarchies within and across a variety of geographic scales, including the international scale. The use of maps to further a group's position with respect to broader debates about Palestine and Israel only makes sense in a context where international actors hold incredible power. This power is effected in part through the legitimacy that technoscientific knowledge provides. Although alternative cartographic methods are becoming increasingly salient, many actors work in the hopes of intervening in international debates where dominant forms of cartography are considered the only true forms. In such cases, it is as if most believe that the master's tools are the only ones. Yet in analyzing the efforts by diverse actors to make the most of their maps, I have attempted to show that the master's tools are hardly monolithic. Still, dominant knowledges are obdurately reinforced across time and space—not infrequently because of the absence of an imagination of any other option. In the context of such reassertions, so much scholarly work involves reappropriating intellectual tools that originally bolstered exclusion. These acts of reappropriation thus link many critical researchers with the empirical cartographers who are the subjects of this book.

Throughout the preceding chapters, I have argued that knowledge and landscapes are fully imbricated. For the Israeli census, PA, and NGOs, methodological decisions are made in concert with knowledge of the kinds of infrastructures that might help or hinder cartographers' travel both locally and beyond. Likewise, the choices of STS scholars are embedded in such material and geographic landscapes, as I contended in the discussion of traveling ethnography in chapter 1, and this involves forms of background-ing that require additional research. So geography cannot be separated from the study of knowledge itself. This is central both in terms of concep-tually mapping variations in paradigms across scales and how practices are situated in particular places as well as alternative forms of movement and situatedness (e.g., figure 6.1).

Figure 6.1
Graffiti on the Wall in Bethlehem depicts Jerusalem in the approximate location where a view of Jerusalem used to be, before the Wall was erected in the early 2000s. Photo by the author.

Nevertheless, if knowledge and landscapes are forever intertwined, then individual case studies are neither entirely local nor fully generalizable. Instead, they require a more thorough understanding of the relations between specific practices and their shapings within variegated geographies as well as scales. Accounts of the geographic production of knowledge, in light of the master's tools, thus serve as a call to attentiveness to geographic boundaries and related exclusions inherent in the practices studied by scholars. It also calls for greater attention to the geographic and material selection and framing of research. Particularly in cases where conflicts and the networks that are part of them exceed the boundaries of any one state (Bank and Van Heur 2007), this requires a more detailed and pragmatic middle-range theory that incorporates multiple senses of the word *middle* (Wyatt and Balmer 2007). These ways that the production of knowledge are inextricably bound up in international landscapes, with their complex overlapping hierarchies that background other people, places, and knowledges, are deserving of much further research.

Writing Other Landscapes: Unbuilding Material Hierarchies of Knowledge

As a medical doctor in rural Egypt in 1956 I asked myself why poor people became more sick than rich people, and I discovered the relationship between poverty and disease. When I asked myself why people became poor I discovered colonialism and dictatorship and politics. ... But what happened when I started to discover the original causes of physical and mental diseases? I lost my job. ... I was placed on the blacklist as a writer and a novelist. In September 1981 I was sent to jail.

—Nawal El Saadawi, *The Nawal El Saadawi Reader*

In addition to internationalism and symmetry, the theme of landscape has involved an exploration of how material hierarchies, including stasis and economic asymmetry, shape the production of knowledge. In the above quotation, El Saadawi describes her transformation from a medical doctor to a scholar and activist who wished to consider the biological causes of disease together with the issue of how capitalist imperialism influenced both mental and physical health. This shift in her thinking brought her into conflict with material hierarchies in Egypt and beyond, resulting in her imprisonment for several months.

The paradigm shift in El Saadawi's thinking was linked to a sea change in her own implicit disciplinary bounds. From being a doctor, she became a novelist and essayist. Her work continues to impress on me the importance of literally (pun intended) pushing the boundaries of the kinds of knowledge that can be produced, expanding not just knowledge itself, but also its material and geographic conditions of possibility.[6] In this book, I have contextualized and specified links between, on the one hand, constructed hierarchies of infrastructures and institutions, and on the other hand, established geographic and economic imbalances. In chapter 5, I explored the method of triangulation that brings forth how knowledge production is situated. To do so, it involves setting the author in dialogue with the research subjects. Such a dialogue is inspired by gender and critical race studies critiques. Rather than imposing an author's critique on the context of study, it might allow the author's own identity to be transformed through the practice of research.

For just as El Saadawi could not ignore the connections between biological health and the diverse conditions of her patients' lives, it is imperative for elite scholars and activists to further examine the economic and geographic conditions of our work as well as the connections between our role as social justice advocates and, for some, simultaneous role as citizens of militaristic governments and scholars in exclusionary institutions. Yet these exclusionary systems of academic practice are also very selectively inclusive. They allow myriad outside scholars, many with economic class privilege, to conduct research in Palestine and Israel, or take up symbolic posts in universities at high pay while temporary educators languish. Through the means of funding and visa controls, however, the same institutions are not as generous when it comes to enabling Palestinians, and to a lesser extent Israelis, control over what kinds of knowledge to make, how to make it, and for whom.

A better understanding of how these forces shape the production of knowledge is also crucial in a context of the broader complicity between universities and the military, including researchers whose methods contribute to the escalation of the occupation. The pairing between the military and academia in particular has increased in intensity over the course of the twentieth and twenty-first centuries (Bowler and Morus 2005, 483), and universities have had a role in the ongoing military cooperation between the Israeli and US governments (Graham 2011). But the relationship between power and technoscience is a complex one, requiring further study of the role of knowledge in broader systems of violence.

In addition to expanding some boundaries, the quotation by El Saadawi also points to the need to be especially attentive to the specificities of broader regimes of complicity, such as when treating her individual patients. Thus, one of my aims in conducting a traveling ethnography was to avoid totalizing characterizations of life in Palestine and Israel. Although the brutality of the occupation continues, nonetheless it is possible for more elite tourists to visit Palestine and Israel, and sit in treelined cafés or lounge on quiet beaches. Indeed, the region remains an important travel destination. Cities like Ramallah and Tel Aviv are widely known for their areas of expensive restaurants and diverse nightclubs that rival those of much larger cities elsewhere. Such privileges—while, for example, Palestinians are killed or sent into prisons that are mere minutes away—can also be seen as part of the occupation.

My attempts at doing justice to the relevant contrasts meant that during my fieldwork, I sometimes went directly from chilly air-conditioned computer labs in Bethlehem and Ramallah, where people in collared polo shirts spoke in low and respectful tones, to tense scenes where young settler children spat in our faces while their mothers threw piles of garbage and human waste on our heads. I alternated between supportive conversations among Palestinian, Israeli, and international aid workers in air-conditioned UN vehicles, and street scenes where teenagers dragged other teenagers across the road by their hair—in the latter case, because teenaged Israeli soldiers were attempting to prevent antioccupation activists from obstructing the settler takeover of a Palestinian family home. So in the spirit of El Saadawi's reconceptualization of her role, I did my best, as a privileged international, to embed myself in the disjunctures that the occupation has wrought. For, every constellation of contrasting expressions of the occupation's power and resistance to that power informs every other. The results, as I have attempted to show here, demonstrate how the material and geographical connections among such seemingly disparate events are not only linked but also have fundamental impacts on the constitution of knowledge.

As a means of investigating these disjunctures, the method of traveling ethnography also helped me to reflexively confront my position as a (perhaps-selectively) privileged international researcher. It helped me to make and remake my role in response to the fact that I was constantly pulled into debates on contrasting sides. While initially I feared that groups would be wary that I might try to speak for them, instead I was often explicitly called on to do so, due to a recognition that my voice might count more abroad precisely because I am a white researcher from Europe and North America.

Of course, I never intended my work to enable me to experience things the way a Palestinian or Israeli person might—a notion that is misguided and even dangerous, besides being impossible. Neither did I attempt to provide a solution, which would easily become a thoroughly patronizing effort. Rather, I hoped to contribute additional understanding from my own selective perspective of how knowledge is occupied. It is occupied both in the particular sense of the control exerted under Israeli occupation and the reflexive sense that knowledge is not almighty. Instead, knowledge emanates from the varying geographical and societal positions that researchers themselves occupy and perform.

Thus, I conducted this traveling ethnography as a way of materializing reflexivity and thereby practicing refraction. It was an attempt to carve out spaces and times in which to be reflexive while undergoing alternative sets of experiences than those of a traditional ethnography, all while helping to foster alternative passages and institutions of knowledge. It was a modest effort to better learn how to build respectful connections of solidarity, even or especially when they were also critical and contentious ones. Nonetheless, I do hope that these conceptions of traveling ethnography and geographic production—practices intended to enable a material cultivation of an attentive disposition—might be useful both in future studies and for the ongoing reconfiguration of scholarly practices and institutions.

New forms of social justice require new institutions, makers, and ways of knowing. Yet these in turn require refashioning the economic asymmetries of knowledge production. While innovative studies on the economics of science have focused on the interrelationships of science and industry (Mackenzie 1990; Scott 1998), here I have used a symmetrical analysis to explore how the content of knowledge is shaped by social, political, and economic injustices within systems of research. They emerged most starkly in the analysis of mobility and stasis in chapter 4, through the description of the raids on PA offices and theft of digital equipment, but they also affect knowledge production and relationships to academia more broadly, as do regional and international restrictions on travel that disproportionately hinder Palestinians. Every visa delay or detention is expensive, not to mention the necessary time and related psychological anguish. These concerns along with the travel required mean that the subjectivities of those whose families cannot direct their human and economic resources toward education are simply framed out of the picture of many researchers and our research.[7]

The economic impacts of international policies are no less significant. Palestine and Israel are areas that have seen a continuous influx of international funding, albeit funds that are unequally distributed and allow varying degrees of control over how they are used. The specifics of such policies, and knowledges they depend on and produce, continue to matter, as do the forms of subjectivity and identities to which they give rise. Studies have demonstrated that at times, international aid may be used simply to shore up the status quo in an effort to stave off the kind of complete humanitarian crisis that could challenge existing regimes (Bouillon 2004; Essex 2008; Hanafi 2005; Hanafi and Tabar 2003; Incite! 2007; Qarmout and Beland 2012; Redfield 2012; Rogers and Ben-David 2008; Tawil-Souri 2006).[8]

In addition, even in the privileged cases where funding abounds, the costs of operating under the occupation are vast. The documents of PA funders mention numerous ongoing delays due to political developments. Furthermore, routine tasks like keeping their electricity running and maintaining their buildings were enormous challenges for many of the organizations where I worked, due in no small part to the particularities of the occupation. The material and economic hierarchies that result when such further costs go underacknowledged can influence research in complex ways. They might alternately discourage or foster the growth of imaginative subjects like El Saadawi, whose definition of health included an analysis of economic inequalities.[9] This begs further study of material exclusion in knowledge production and the reproduction of hierarchies through elite forms of knowledge.

Permanent Visas for Traveling Cartographers

Burnout and despair can be the lot of anyone concerned with social justice, but they must be balanced against the consequences of a lack of concern.

Indeed, El Saadawi (1997) also remarked as much in the dedication to her reader: "To the women and men who choose to pay the price and be free rather than continue to pay the price of slavery." I wrote this book in part to thank all those with whom I worked in the region, particularly those who stubbornly refuse to be governed by fear, to not conflate what has happened thus far with what could happen in the future, and maintain that the personal price of moving against the current is worth the new channels that are worn down as a result.

Nonetheless, despair and burnout are ever present. As I write this, people continue to die in the streets. Commenters are beginning to talk of a Third Intifada against the Israeli occupation, even as the Israeli government ponders closing off East Jerusalem entirely in yet another form of collective punishment. On the one hand, the occupation seems to be brutally clear, and its termination inevitable. On the other hand, there appears to be no simple formula to reach an outcome whose time has come, and with each day we wait, the occupation continues to entail so much suffering. Yet notions like *formula* and *solution* can themselves be part of the problem. Even knowing this, I also am aware that researchers are sometimes locked into doing things the same ways, in ways that further colonial legacies, despite our best efforts. Nonetheless, I do believe that we can be smarter and more reflexive than we have been so far.

Acknowledging the very real difficulties of breaking out of current material pathways for empirical knowledge is part of working for viable alternatives. For example, figure 6.1 illustrates the creativity of making alternative views, as in the graffiti that re-creates a view of Jerusalem that is currently blocked by the Wall.[10] Yet it also illustrates the challenges of dismantling that Wall, which would reinstate a less obstructed view across the landscape. Innovating new passages thus requires both reimagining and remaking the master's tools. Darwish acknowledged this repeatedly in his work, where he wrote of how land and ideas can reinforce one another through maps. Speaking of a "land without borders" Darwish (2007, 205) notes, "When we walk in its map it becomes narrow with us." He thus contrasts the land, which is without borders, with the more dominant views in maps. Because walking not in the land but instead in its *map* is what is narrow. At the same time, however, narrowness might not need to be seen only in a negative light, as it also reflects the particularities of the walk and those doing the walking.

Such different ideas of landscape, in ways that critique dominant conceptions of sovereignty and territory, can also help to produce alternative geographies of knowledge. These in turn can help to generate ever more varied epistemologies and theories through the alternative passages for knowledge that they help to create—passages that might be both narrow and wide at the same time, depending upon the perspective from which they are viewed. Despite the difficulties evident in the cases here, it is not completely impossible that dominant knowledge forms might yet be dismantled and remade, with a little creativity and a fuller understanding of the spatial and material challenges at work. Their parts then might be ingeniously recombined and remolded such that although not entirely unrecognizable, they bear little resemblance to the original. Restrictive legacies might be reshaped and recirculated into practices, dispositions, and landscapes that are refreshingly unexpected, and helpful for wholly new purposes, the likes of which have yet to be imagined or understood.

Notes

Chapter 1: Where Cartographies Collide

1. For some examples of GIS maps, see figures 2.1, 3.4, 4.2, and 5.5a.

2. For a paraphrase of historian Dipesh Chakrabarty on how concepts can always potentially translate across contexts, see T. Mitchell 2002, 7.

3. I focus throughout on publicly available maps. I was prevented from studying currently classified maps for the simple reason that I had no access to them whatsoever. Since the overall concern was with maps intended to publicly influence scientific and political debates, however, a lack of access to secret maps was less of an obstacle than might be expected. So this book only addresses military cartographic technologies and maps in cases where they have been adapted for unclassified use. For an attempt to grapple with the liminal character of classified knowledge, see Rappert 2009. For an analysis of the impact of censorship and classified knowledge on academia in Israel, see Forte 2003, and for a recent plan to effectively reclassify hundreds of millions of documents in the Israel State Archives, see Matar 2016; Qato 2016.

4. Although Marston, John P. Jones, and Keith Woodward (2005) argue for doing away with the notion of scale, it is so integral to the present subject that it was necessary to include it in the analysis. Indeed, the instrumental use of scale is one of the mechanisms of the occupation. For example, the term *Israel* might implicitly include the Palestinian Territories when it is expedient for the speaker to do so (such as when seeking to claim land for the Israeli state), but exclude them when it is not beneficial for the speaker (such as when debating the state's obligation to provide access to education, housing, and work).

5. The boycott is increasingly felt in international academic circles. This was seen, for example, in the extensive discussion on the critical geography e-mail list CRIT-GEOG-FORUM regarding whether international academics should attend the 2010 International Geographical Union annual conference in Tel Aviv. In addition, the International Critical Geography Group chose to hold its 2015 conference in

Ramallah. During the conference, its participants readily passed a political statement that strongly endorsed the PACBI as well as supported the wider Boycott, Divestment, and Sanctions (BDS) Movement with respect to Israel. The boycott is also filtering into academic publishing. Zone Books issued a recent edited volume, *The Power of Inclusive Exclusion*, with a thin slip of printer paper in the front cover noting that "although the research group which was the genesis of this publication was funded by the [Israeli] Van Leer Jerusalem Institute ... the publisher of this volume ... received no funding or other support from the Van Leer Jerusalem Institute, nor is Zone Books in any way associated with that institution" (Ophir, Givoni, and Hanafi 2009).

6. These might include Palestinian citizens of Israel—who, by Israeli state accounts, formed over 20 percent of Israel's population in 2012 (CBS 2012)—and those of mixed Palestinian and/or Israeli heritage as well as Mizrahi Jews, or Jews of Middle Eastern origin who are reported to make up 50 percent of the Jewish population of Israel (Loolwa 2013). Mizrahi Jews are often citizens of Israel, yet they have been discriminated against in a context dominated by European, Ashkenazi Jews. In 1971, the Israeli Black Panther Party was formed as a response to the treatment of Mizrahi Jews (Ettinger 2007; Israeli Left Archive n.d.). All these groups also overlap in ways that are ignored by the simplistic dichotomy between Palestinian and Israeli identities, which is rigidly imposed under the occupation.

7. For an investigation of the broader context of technology and empire both within and beyond Palestine and Israel, see chapter 2.

8. Elif Shafak (2005) has given the name "linguistic cleansing" to a similar process that took place in Ottoman Turkey. For Mahmoud Darwish's description of his partial estrangement from this brutally altered landscape, see chapter 6.

9. For an Israeli account of the history and broader context of Israeli society both before and during the 1967 war, see Segev 2007.

10. The terminology in Palestine and Israel is complicated, but it is also incredibly important. The subject deserves—and has—its own books. Terms serve to lay out the political positions of a speaker even before an argument has finished being uttered. For practical reasons alone, I use *Occupied Palestinian Territories* (OPT) and *Palestinian Territories* interchangeably to refer to the West Bank, the Gaza Strip, East Jerusalem, and the Golan Heights, among other smaller areas occupied by Israel in 1967. (Israel likewise occupied the Sinai Peninsula in 1967, but the Sinai was turned back over to Egypt in a process that ended in 1982.) I use *Israel and Palestine* or *Palestine and Israel* equally to refer to the combined areas of the OPT and Israel. In some instances, *Israel/Palestine* appears to lend support to one binational state, a one-state position, while *Israel and Palestine* would be seen to support separate states of Israel (formally annexed Israel) and Palestine (in the OPT), a two-state position. I alternate the latter for clarity and variety, though, rather than to claim one overall position. In the context of the lack of independence and control granted

to Palestinian institutions, more and more critical scholars are supporting the one-state solution. The one- and two-state debates are also briefly discussed in chapter 5, although the full debates fall outside the scope of this book.

By contrast, some authors use *Occupied Palestine* to refer to all of Palestine and Israel, given that the founding of Israel after 1948 also was effected through the occupation of Palestinian land. In contrast, others use *Israel* to refer to all of Palestine and Israel as well as show support for Israel's (in my view, horrendous) occupation. My choice of terms of course has political consequences of its own. I try to own these to the best of my ability. Instead of a definitive statement, it is meant as an imperfect but necessary strategy for conducting a geographic study that, without the use of some terms to designate the areas discussed, would be impossible.

11. As discussed in chapter 2, they also helped to further develop techniques of imperial rule through the use of infrastructure systems like roads and bridges (Salamanca 2014).

12. Part of the challenge of studying quantitative cartographers from the mid- to late twentieth century is that they often do not speak at length on their methods, even though many of them repeatedly and thoughtfully revised those methods. Such things were expected to be entirely clear and apparent, and therefore not worthy of discussion.

13. For a discussion of the Israeli confiscation of Palestinian books and maps, see chapter 4.

14. For an investigation of the making of facts, see chapter 5.

15. Michael Fischbach (2003) suggests that the request for restitution was simply an excuse, and that political considerations were in fact behind Hadawi's resignation. These would have included outright discrimination—namely, opposition to having a Palestinian on the team. It is likely that Hadawi did not foresee even the perception of a conflict of interest or else he never would have submitted the request.

16. My interviews were recorded and conducted for the most part in English, with portions and interjections in Arabic and Hebrew. During the participant observation, I took part in different meetings and tours that were conducted either in Arabic, Hebrew, and English, or a combination of all three.

17. Another tactic that I witnessed many times at checkpoints was to temporarily delay entry from the West Bank into Israel in an apparent effort to cause mental and emotional anguish—partly, it seemed, out of a mixture of boredom and anger on the part of the guards. The checkpoints have full-body barred turnstiles of the sort that might be seen in a sports arena or public transit stop. Even children must pass through these alone. In one case, a Palestinian boy who was about eight years old and wearing a Santa suit was with a woman who might have been his grandmother. They tried to pass into the turnstile together, but the guards separated them. She then tried to allow the boy through first, but the guards forced her to go

before him. Subsequently, when he was in the middle of the turnstile, prevented by the bars from either going forward or backward, they locked it and held it that way for twelve minutes—a period I timed by using my phone. Twelve minutes might not seem like a long time to wait, unless you're a young boy in a Santa suit at a military checkpoint, separated by bars from your grandmother on one side and a long line of people on the other—others who are watching you impatiently because they're waiting as well—while two guards point their machine guns impassively at your head, and a third looks on. After this unexplained pause, the turnstile was unlocked, and the woman and boy proceeded through the checkpoint without further incident.

18. My ability even to conduct research full time for an extended period was due to a generous academic position at Maastricht University.

19. It therefore resonates with contemporary feminist and postcolonial theory more broadly, including Sara Ahmed's (2006, 2010, 2014) perceptive and influential conceptions of *orientations* and *willfulness*.

20. As discussed in chapter 2, the work of Bruno Latour, a key figure in the development of Actor-Network Theory, includes extensive references to maps and geography. For a further elaboration of the notion of space as background and related practices of *backgrounding*, see chapter 6.

21. On the notion of *passages*, see Peters, Kloppenburg, and Wyatt 2010.

22. Thus while sympathetic to the role of nomadism in critical theory (for a prominent example, see Braidotti 2013, 57), by reappropriating the concept of *landscape*, I seek to emphasize that nomadism is not simply aimless mobility or some unobstructed form of wandering around. Instead, nomadism encompasses a variety of patterns of daily life that require detailed knowledge—including technical knowledge related to botany, navigation, biology, and astronomy, among many others—and that meet enduring resistance.

23. See the emerging literature on alternative tourism in Palestine and Israel (Barnard 2013; Stein 1998), which includes critical analyses of precisely what constitutes "alternative." Nonetheless, numerous international tourists from around the world visit the West Bank each year, and their presence supports a key industry in an economy that is heavily dampened due to the occupation. The most recent Lonely Planet guide for Israel and the Palestinian Territories provides detailed information on travel to the West Bank, and several illuminating alternative travel guides have been released in recent years (Jubran, Nasser, and Yahya 2008; Natour and Abu Ta'ah 2000; Shahin 2005; Szepesi 2012).

24. On issues facing overresearched communities, see Sukarieh and Tannock 2012.

Chapter 2: The Materiality of Theory

1. The Wall is known by many names, which have varied political connotations. In English, they include the Separation Wall, Segregation Barrier, Apartheid Wall, and (as featured on the Web site of the Israeli Ministry of Foreign Affairs) Security Fence.

2. On the effects of the Wall on divided communities, also see Heruti-Sover 2012; Hammerman 2011.

3. On recent surveillance in Palestine, see Tawil-Souri 2011c; Halabi 2011. On the role of technology in the enclosure of the Gaza Strip, see Tawil-Souri 2011a.

4. Critical GIS is related to, but distinct from, critical literatures in the disciplines of architecture and urban planning.

5. It should be added that if everyone knows, then it is because they belong to a shared community where meaning is attributed to such statements as "This farm has been in our family for many generations."

6. In Mandate Palestine, the British argued that they furthered geopolitical and humanitarian goals. However, their focus on standardizing property ownership belied their (at times conflicting and variable) colonial motives (El-Eini 2006; Gavish 2005).

7. Mitchell argues that the boulevards resulted from a wider belief in structure, rather than indicating that cities are naturally structured according to universal rules, as is commonly assumed. Yet the wide boulevards of cities would also facilitate later mapping and surveillance, thereby reinforcing the apparent effectiveness and necessity of the structure. So even if a belief in structure led to the construction of structured landscapes, those produced landscapes subsequently affect knowledge production in ways that exceed that belief.

8. For a deconstruction of the concept of the *Middle East* using maps, see Culcasi 2008, 2010, 2012).

9. For the relationship of Lefebvre's work to French research on science and social space in the context of aerial photography, see Haffner 2013; Merrifield 2000.

10. Mitchell's claim (2002, 79) that Lefebvre assumes a distinction between the social and material has been countered in part by work in geography that adapts Lefebvre's views to the production of nature (e.g., C. Katz and Kirby 1991; Smith 2008; Swyngedouw 2006).

11. See also Rafi Segal and Weizman's (2003a) coedited volume, *A Civilian Occupation*. The Association of Israeli Architects censored the book's first edition (Segal and Weizman 2003b), and attempts were made to halt the construction of the new

Israeli National library, designed by Segal, in relation to his continued support for the book (Rosenblum 2012).

12. In a related vein, Clark, Massey, and Sarre (2008, 2) pair the study of land with mobility, arguing for combined analyses of territory and flow, which I explore more fully in chapter 4. See also Ingold 2007, 2011.

13. See, for example, the analysis in chapter 5 of settlement cartography from *within* and *without*.

14. Hadawi (1957, 7) perhaps ironically referred to this as an effort to turn the desert into a "rose garden." The often-repeated adage of a "blooming" desert itself relies on mistaken colonial notions that deserts by themselves are unfertile and economically subordinate to the supposedly more productive landscapes of northern Europe. On how, through a combination of agricultural science, cultivation practices, and ecological forces, the "drylands" became "a terrain of technopolitical action," see Tesdell 2015.

15. The challenges faced by the PA are dealt with in chapter 4. For an example of an attempt to combine modernization with environmental sustainability, see Saleh 2008.

16. On the differences in the treatment of refugees and nonrefugees in the urban planning of Ramallah, see Alkhalili 2012. On development and public space in Amman, Jordan, see Coignet 2009. On gender and public space in the OPT, see Abbas and Van Heur 2013; Zawawi, Corijn, and Van Heur 2013.

17. For a further elaboration of specific European colonial tropes with respect to development and technology in the global South, see Adas 1989.

18. Linear conceptions of development and progress can be traced in part to the work of the fourteenth-century historian Abd al-Rahman Ibn Khaldun. Ibn Khaldun's writings were widely influential in the Arab world, and by the eighteenth and nineteenth centuries, in Europe. In the introduction to his *Muqaddimah*, Ibn Khaldun argued that a breakdown in social cohesion could lead a society to be overtaken by outsiders from its geographic periphery. Although Ibn Khaldun can be seen as supporting a linear development of history, this is only one interpretation. Indeed, his processual view is more in keeping with that advocated by Thomas Kuhn (1996), who famously asserted that science consists of a succession of different paradigms whose proponents vie for dominance based on collective support and belief. For an English translation of the *Muqaddimah*, see Ibn Khaldun 1967.

19. See also Lowdermilk 1960.

20. In turn, colonial tropes in the Americas were themselves built on forms of Orientalism, based on the denigration of Muslims and Jews alike, that were brought over with Christopher Columbus during his conquest (Shohat 2002). In recent years, Orientalism has likewise been mobilized in attempts to enforce strict

dichotomies between East and West, North and South, as well as between Israeli and Palestinian academics living in Europe and North America (Shohat 2004).

21. At times, colonial development discourses persist with surprisingly little variability. To provide a recent illustration, in August 2013 on the progressive blog *Mondoweiss*, one commenter criticized an article against Israeli settlements by claiming, apparently without irony, that "Israel is a developed nation in a sea of economic backwardness" (Robbins and Weiss 2013).

22. Even critical scholars such as Kurt Goering (1979), although sympathetic to the impacts of development on Bedouin communities, nonetheless sometimes perpetuate colonial tropes by mistakenly presenting Bedouin societies as timeless and unchanging (Fabian 2002). This shows how difficult it can be, even for professional researchers who work in relevant areas, to move past ways of thinking that are rooted in colonialism.

23. Variations on arguments about security and technology certainly are not confined to contemporary Israel. To give just one example, Wolfgang Schivelbusch (1987) demonstrates how in Paris, safety concerns were used to legitimate the introduction of electric streetlights in the nineteenth century. At the time, those who feared that streetlamps enabled greater police surveillance of the public would routinely smash them.

24. For more on present-absentees, see Nur Masalha's (2005) edited book on internal refugees. Masalha's own chapter in the same volume is particularly relevant. It outlines the development of and resistance to brutal Israeli land policies, including the notable case of the villages of Kafr Bir'im and Iqrit near the Lebanon border as well as the broad Israeli crackdown on Palestinian public remembrances of 1948.

25. Geopolitical motives were also founded on racist colonial ideologies (Massad 2003, 2006). For example, they mistakenly implied that only those of European descent are capable of being rational and democratic, and that narrow notions of rationality and liberal democracy were the only possible means of positive social organization.

26. Not unlike contemporary Palestinian efforts, Israeli developers sought to achieve some unity of local (natural) conditions and rationalized construction. Recently, environmental movements that seek to mitigate the detrimental effects of land rationalization have arisen in the region. Such groups, however, must also actively deal with the politics of conflict and occupation (Cohen 1993, 1994, 2002). In this context, for example, *sustainability* has become a loaded term. To the PA, sustainability is used to refer to attempts to prevent the destruction of Palestinian areas in the West Bank under the occupation, in combination with programs to reverse the negative impacts of past modernization. In contrast, among some environmental groups in Israel, sustainability is used in the context of maintaining an

explicitly Jewish presence on the land (Cohen 1993, 1994, 2002). Environmental conservationism and nationalist efforts thus can readily overlap. On the use of national parks for political ends, see Hasson 2013; Zonszein 2013a.

27. Two exceptions of work on colonial knowledges that incorporate geographic and STS critiques are Nadia Abu El-Haj's (2001) pathbreaking study of the building of Israeli archaeology as a field discipline as well as Gary Fields's (2010, 2012) comparisons of land enclosure movements in the historical United States and United Kingdom with landgrabs in contemporary Palestine.

28. This relates to the geographic shaping of subjectivities more broadly—a theme that Wood (2010) treats in his analysis of artistic maps. Wood also explores the implications of artistic atlases. These include *Atlas of the Conflict* (Shoshan 2010), which combines empirical maps and design, as well as the *Subjective Atlas of Palestine* (de Vet 2007) and recent exhibit catalogue *Imaginary Coordinates* (Rosen 2008).

29. Not coincidentally, *wadi*, a term for a type of river-flooded valley, comes to Modern Hebrew from Arabic. Also from Arabic, it survives in Spanish in the prefix *guadal-*, as in *Guadalcanal*, which comes from the Arabic name *Wadi al-Qanal*, meaning "valley of the stalls."

30. The hype that once surrounded large-scale industry and agriculture in Israel has transferred to high-tech companies. See Senor and Singer 2009.

31. For an analysis of the alternate highlighting and omission of the location of Israeli military bases on maps, one example of the different ways one highly politicized type of feature is incorporated into cartography, see chapter 5.

32. Chapter 3 focuses on the work of one individual, Roberto Bachi, but the goal is precisely to place his research in its broader political context instead of solely elaborating on his personal motives.

33. Definitions of writing are also being expanded to include the phenomenological and material aspects of the act of writing. Chaim Noy (2008) has asserted as much in the case of the ideology of commemoration in Israeli historical sites.

34. More broadly, new research is emerging on the Anthropocene and indigenous spatialities. This includes anthropological studies such as Elizabeth Povinelli's (2006) work on *geontology*. In a move that resonates strongly with debates in Actor-Network Theory and critical geography, Povinelli analyzes Australian aboriginal conceptions of the subject that seamlessly incorporate both the earth and human body. Yet her emphasis is on specific instances of place making more than on the relationship between such aboriginal conceptions, or the economically as well as geographically situated methods that she herself uses as an international scholar who produces textual knowledge—the latter of which is the focus of public participation GIS research.

35. Latour clearly acknowledges the ways that geographic space is social, in a way that is in sympathy with critical geography. He focuses less on one crucial effect of social space, however: that spaces reflect imbalances of power that are not easily traversed by all, or even most, researchers. Latour aptly critiques simplistic ideas of nested geographic scales, claiming that "macro no longer describes a wider or a larger site in which the micro would be embedded like some Russian Matryoshka doll." Instead, as many geographers have argued, macro is also "another equally local, equally micro place, which is connected to many others" (2005, 176). Yet the power dynamics of these sorts of macro–micro connections and interweavings differently shape the type of "plodding along" through space that the researcher is called to do.

Chapter 3: Removing Borders, Erasing Palestinians

1. For a digital version of the 1967 census data for the Palestinian Territories, see Perelmann 2012.

2. On representation and maps, see chapter 2.

3. A census also figures prominently in the biblical story of the birth of Jesus, where it is given as the reason why Jesus ended up being born in a stable. Mary and Joseph are said to have traveled to Bethlehem to be counted in a Roman census, and the reason they couldn't find a place at an inn was because so many others were in town for the census as well.

4. Zureik's work also contributes to the large body of literature on Palestinian resistance to these asymmetrical relations, which I discuss in chapter 6.

5. The persistence of anti-Semitism, or racist prejudice against Jews, is well documented in a variety of contemporary contexts. In recent years, however, there has been a campaign on the part of advocates of Israel to conflate any criticism of Israel or Zionism with anti-Semitism. In effect, it involves accusing someone of anti-Semitism any time they draw attention to an oppressive policy enacted by the state of Israel. Pro-Israeli settler movements use this tactic in an attempt to close down dialogue and deflect attention away from the ongoing racism directed against Palestinians, who are systematically oppressed under the Israeli occupation. Although largely unsuccessful, such strategically misguided instances of "crying wolf" do have the unfortunate side effect of hampering acknowledgment of instances of anti-Semitism when they do occur. On racism against Palestinians, see Joseph Andoni Massad (2003, 2006), who provocatively argues that anti-Palestinian racism is a form of anti-Semitism.

6. In chapters 4 and 5, I explore the implications of knowledge hierarchies for Palestinian cartographers.

7. Anne Godlewska (1999) has pointed out how in the Middle Ages, geography was also viewed as a science, and that it was founded on accurate descriptions rather than hypothesis testing. It lost this status as a science during the Renaissance, only to regain it in the twentieth century. In addition, Godlewska convincingly investigates exceptions to the broad trends described here—the explorer and cartographer Alexander von Humboldt foremost among them.

8. For more on the early history of GIS use in the census, see the GIS History Project 1997.

9. This position is still put forth, as can be seen from the claims of several candidates in the 2012 US Republican primary and numerous online articles, such as Joseph Farah (2002). Also, the online "Map of Palestine" (2015) directs viewers to a page that says, "The map you requested could not be found. It is possible that the address is incorrect, or that THE COUNTRY DOES NOT EXIST!"

10. Meir's words also contrast with a well-known quote from Angela Davis (1971): "Human beings cannot be willed and molded into nonexistence." Davis was writing from prison just two years after Meir spoke, albeit in the different context of the US Black Panther struggle. Viewed one way, Davis's quote seems to insist on an essentialist materialism where physical presence cannot be manipulated or constructed through social means. Yet her statement instead could be read pragmatically—that is, to suggest that human beings will not often *allow* themselves to be willed into nonexistence. In this respect, Davis's words highlight how attempts like Meir's can actually be counterproductive, because they frequently generate the types of resistance they seek to foreclose.

11. In a related effort, PLO and other Palestinian maps did not show the Green Line even before 1967 (e.g., Hadawi 1957) although as with Israeli maps, this practice has mitigated somewhat over time.

12. This is epitomized in the disastrous category of *present-absentees*, as noted in chapter 2.

13. In chapter 4, I focus on the Palestinian state mapmaking efforts that began during this period.

14. The snapshot metaphor was built on the census methods under the British occupation. For example, Hadawi (1957, 8) argues that the 1931 census "enumerated all the persons present in Palestine at midnight on November 18th, 1931, irrespective of whether they were residents of the country or not." Similarly, according to Hadawi, residents who were not in Palestine at the time were not enumerated, and this held for further estimates, based on the census, right up to 1946.

15. More recently, Google Street View, which is described in chapter 1, has begun to collect data for the Israeli settlements (Sheizaf 2012). Street View has not been extended to the Palestinian Territories more broadly at the time of writing (see figure 4.5b).

Chapter 4: The Colonizer in the Computer

1. On internationalism in Palestinian political activism, see Khalili 2007.

2. This chapter further develops and extends my previously published article, Bier 2017.

3. *Sumud* is pronounced with a hard "s" and two long vowel sounds similar to the "oo" in the English word *food*. It is alternately transliterated as *soumoud*.

4. Respectively, they can be reached via the following links: https://twitter.com/SumudHouse (accessed December 9, 2015); http://www.aeicenter.org (accessed December 9, 2015).

5. On urban obduracy, see Hommels 2005. On backgrounding, see chapter 6.

6. On Palestinian immobility and the Internet, see Aouragh 2011a, 2011b; Tawil-Souri and Aouragh 2014. For an analysis of the role of North–South hierarchies in the movement of technology and critique of simplistic notions of technology transfer, see Shamsavari 2007; Edgerton 2007.

7. For example, on Shehadeh's ultimate rejection of a PLO post due to the dominance of the Tunis faction, see Said 2001, 151–152.

8. In addition, there were differences between the pre-2007 Fatah in the Gaza Strip and the Fatah-backed PA in the West Bank, and Hamas remained an unofficial player in the West Bank.

9. For this reason, more recent treaties often use lists of coordinates instead of maps as the definitive indications of contested borders (e.g., Hashemite Kingdom of Jordan and State of Israel 1998). Yet there are also numerous issues with determining the framework and position of the coordinates to be used.

10. Hadawi (1957, 25) has argued that the area of such villages "could not be calculated accurately because the Arab military authorities would not permit, for security reasons, either the government surveyors or the villagers to approach within 500 yards."

11. This also contrasts with local and regional mobilities, however, as described in chapter 1. Transit within as well as between the Gaza Strip and West Bank is highly restricted, as is movement from the Palestinian Territories into incorporated Israel. In addition, at times the Egyptian government closes the Rafah border crossing between Egypt and the Gaza Strip.

12. For an example survey of environmental impacts in Tulkarem that makes use of current maps and fieldwork to make governance recommendations, see Dweib 2011.

13. On PA surveillance, see Parsons 2011.

14. Certainly there are also significant differences between the British and Israeli regimes and their maps. But what is most notable here is that disparities in knowledge are reproduced precisely when it comes to attempts to control local populations.

15. The same report points out that cartographic fieldwork "was less seriously affected as most of the surveyors were Arabs and working in Arab areas." The survey efforts were eventually abandoned, however, due to the onset of war (Mitchell 1948).

16. In a different context, Latour and Emilie Hermant (1998) note that a 1930s engraved panorama in the Samaritaine department store in Paris has outlived many of the buildings that it depicts.

17. To my knowledge, there has been no effort by any party to enforce a copyright on Mandate maps in Palestine and Israel, and they are widely available for purchase on the Internet. Access to such maps, however, is often filtered through the restrictions of the holding institution, such as the British Library or Palestine Exploration Fund archives.

18. Speaking of the development of digital cartography, Laura Kurgan (2013, 17–18) has noted a similar process of the increasing "'coordination' of physical space with digital space" that parallels, to some degree, the self-fulfilling prophecy described here. However, here I am concerned with the ways digital and physical landscapes are fully imbricated from the beginning.

19. For examples of both historic and more contemporary aerial photographs of the region, see Kedar 1999.

20. Abu-Libdeh has since been brought up on corruption charges—charges, however, that appear to be unrelated to his time as the director of the PCBS, and that he claims are politically motivated ("Palestinian Economy Minister" 2011).

21. On the impact of political events on development funding in Palestine, see also Daoud 2000. Shehadeh (2013) has suggested that PA authorities and academics may become discouraged, after years or even decades of trying, due to the high personal and economic costs of doing research.

22. The difficulties of maintaining archives in colonial and postcolonial contexts has led Craig Buckley and Mark Wasiuta (2013) to advocate privatizing archives. As I show here, however, the particularization and dispersal of sources, while they might solve some problems, simultaneously introduce an entirely new set of concerns. Additionally, there have been some efforts to institutionalize research in the absence of a recognized state, including the creation of the Palestine Academy for Science and Technology. But this is not to paint too rosy a picture of the strength of institutions in Europe and North America. For example, Wiebe Bijker, Roland Bal, and Ruud Hendriks (2009) have referred to the "paradox" of the trend in the

Netherlands toward the increasing authority of scientific advice at the same time as institutions of science are being eroded. With thanks to Elena Glasberg for bringing Buckley and Wasiuta's work to my attention.

23. The withholding of data on the part of the Israeli authorities as well as the classification of many maps and related documents also led to the creation of the PA mapmaking projects to deal with both the content and availability of data. Classification creates high levels of uncertainty and continues to shape research. As noted in chapter 1, in 2016, the announcement of a supposedly "benign digitization project" at the Israeli State Archives revealed that it would involve the effective reclassification of hundreds of millions of historical documents (Matar 2016). In some cases, though, depending on time and funding, the differing levels of security across multiple collections can be used to obtain a more thorough conception of the maps produced. Given this, it is possible to work within different Palestinian family archives (Qato 2016); compare Turkish, Jordanian, Israeli, and British collections; or use library sources in combination with those available from private retailers.

24. The funding provided by supranational organizations like the European Union and donor countries also significantly shaped PA cartography in ways that certainly deserve further study. For example, the Norwegian Agency for Development Cooperation granted extensive funding to the PPIB, but the sometimes-minute monitoring of those funds could be seen as a form of imperial control.

25. Although not a Fulbright fellow, one student from Gaza did attend Oxford, with her education paid for in part by fellow students and activists at the university who built up a fund through numerous small donations of about 4.50 euros each ("Student's Journey" 2013).

26. It also contrasts with efforts to expand the already-high mobility of Israeli elites. Locally, these include the construction of transportation infrastructure to serve settlements in the West Bank ("Israel Unveils Controversial Plan" 2013). Internationally there are attempts on the part of US senators to lobby the US government for a visa waiver exemption for Israelis who travel to the United States. This is surprising because contrary to standard policy, the Israeli government is not prepared to offer parallel waivers to US citizens traveling to Israel (Coogan 2013).

27. In a similar vein, at the first TEDx Ramallah, Laila Atshan (2011, 1:40), a counselor based in Ramallah, noted her extreme admiration for Palestinian taxi drivers, to extended applause.

28. To a lesser degree, such challenges have been faced by Israelis seeking to carry out alternative cartographic fieldwork in the West Bank, even before the increased restrictions on travel outside settlements. In the 1980s, Meron Benvenisti and Shlomo Khayat (1988, n.p.) could assert, in the introduction to the *West Bank and Gaza Atlas*, "We have done the best we could, given our very limited resources and less than satisfactory cooperation rendered by the [Israeli] authorities."

29. The coast was also long the focus of Palestinian memories of loss after 1948, as Sandra Arnold Scham notes (2003, 75). Despite their broad scale, the British topographic maps still are invaluable repositories of information on the number and location of Palestinian villages before 1948. They continue to be referred to as factual sources—for example, in a recent Facebook post by the antioccupation group Ta'ayush. The author of the post uses images of pre-1948 British maps to argue that the Israeli military "cannot be persuaded with the facts." In this case, the facts are the existence of villages in an area where the authorities seek to make a military firing zone, which requires forcibly emptying it of (Palestinian) civilians. See https://www.facebook.com/170738066985/photos/a.196756516985.138990 .170738066985/10151551881596986/?type=3&theater (accessed February 2, 2017).

30. For a fuller account of the impact of uncertainty on specific visualizations and digital models, see Kouw, van den Heuvel, and Scharnhorst 2013.

31. Even prior to 2006, the focus tended to be more on the West Bank than on the Gaza Strip.

Chapter 5: Validating Segregated Observers

1. Mathematically, 86.4 percent and 0.54 percent are *already* percentages, so it would be common to simply use the error rather than the percent error. To find the error, one would simply subtract: 86.4 percent − 0.54 percent = 85.86 percent (error). Instead, the author determines the percent error of the percentages: (86.4 percent − 0.54 percent) / 0.54 percent * 100 = 15,900 percent (percent error). Although both are technically correct, by comparison the figure *15,900* percent was likely chosen because of its rhetorical effect. Such assertions of error were widely posted on pro-settlement Websites, and even filtered without citation into a handbook for the Israel lobby in the United States (International Business Publications 2011, 32). Yet this line of criticism doesn't even include those who argue solely using biblical sources for land ownership (e.g., Wild Olive, n.d.).

2. ARIJ's Arabic name can be transliterated as *Ma'ahad al-Abhath al-Tatbiqiyat al-Quds (Areej)*. Peace Now's Hebrew name is most often transliterated as *Shalom Achshav*. For both organizations, the English names are literal translations of the terms. ARIJ is said as a word rather than an acronym. The *a* is short as in the word *talk*, the *i* sound is long as in *we*, and the *j* sound is pronounced as in the word *jeans*.

3. Israeli settlements are often positioned close to existing Palestinian towns and given names that are similar to the existing towns' names. This is intended both to maintain (or instate) geographic place markers that are said to have biblical roots and serve as a reminder of the settlers' goal of forcibly replacing the Palestinian towns. In Arabic, Taqu'a is written تقوع. Tekoa, a nearby Israeli settlement, uses the Hebrew version of the name, which is written תקוע. There is no common English transliteration for either term at the time of this writing, and I simply use the most

common spelling. Several other spellings, such as *Tequ'* and *Tuqua*, are in regular use as well.

4. Overall, Wolfson's response centers more on an apparent attempt to discredit ARIJ by making stylistically exaggerated claims rather than providing a substantial analysis. To give just one example, Wolfson omits one key detail of the ARIJ report. The report doesn't assert that the military had already cleared the trees. Instead, it maintains that the military announced that it would begin clearing them. So from the outset, Wolfson is attempting to counter a claim that doesn't actually appear in the ARIJ report.

5. The moment that NGO Monitor strayed from such (allegedly impartial) empirical criticisms was precisely when it was convicted of breaking Israeli libel laws. In 2007, after spuriously claiming that ARIJ and other groups "emphasize external issues including the justification of violence," ARIJ and several partner NGOs sued NGO Monitor for libel in Israeli courts, and won. As a result, Professor Gerald Steinberg, the executive director of NGO Monitor, was forced to issue an apology (Silverstein 2010).

6. This chapter is the result of five months of participant observation in the GIS unit at ARIJ in spring 2011 as well as ongoing interviews and participant observation during that same period at Peace Now. It also includes insights from my interviews with the major cartographic organizations in the region, including those that identify alternately as Palestinian, Israeli, international, or joint (cooperative) NGOs as well as relevant organizations within the United Nations.

7. The debates about facts spill over into social media in self-referential ways. This can be seen in the following quote from the English Wikipedia page "NGO Monitor" (2015): "The online communications editor of NGO Monitor, Arnie Draiman, was indefinitely banned from editing articles about the Israeli–Arab conflict for biased editing, concealing his place of work and using a second account in a way that is forbidden by Wikipedia policy. Draiman was a major contributor to the articles of his employers ... and performed hundreds of edits of [those] human rights organizations ... to which NGO Monitor's president, Professor Gerald Steinberg, is opposed." The talk page was only available in English, so these assertions were absent from the Hebrew, Russian, and Dutch versions of the "NGO Monitor" page.

8. Scholars in the growing field of settler colonial studies are pursuing the study of Israeli settlements. As argued by Omar Jabary Salamanca, Mezna Qato, Kareem Rabie, and Sobhi Samour (2012), their work shares with this book a focus on how Palestine and Israel are *not* exceptions or isolated solitary cases, the incorporation of mezzo-level theory, and a commitment to study the Israeli occupation as a phenomenon that is fully imbricated in broader social, political, and material worlds. As a contribution to this literature, I add an analysis of the efforts of different researchers in varied contexts to produce knowledge of settler colonialism

with respect to their positions within settler landscapes. On segregation in Jerusalem, see Pullan 2013.

9. For an example of a Palestinian journey that would be manifestly different for non-Palestinian Israelis or internationals, see Yousef 2011. On the effects of segregation among Palestinians from different areas, see, for example, Ghussein and Robbins 2013.

10. For an investigation of how imaginations of social distance between groups have been transformed into measurable, statistical distance through graphs and other visualizations of migration, see Boersma and Schinkel 2015.

11. Both this chapter and chapter 4 study the relationship between power, landscapes, and mobility. Yet chapter 4 focuses on the implications of how institutions and infrastructure shape maps at the level of an entire data set, such as the data for West Bank roads, whereas this chapter analyzes the specific decisions made in the mapping of individual points and buildings.

12. It was seldom asked whether science or facts provide the best way to require someone to report the demolition of their home, or the imprisonment or killing of their family.

13. On the lack of acknowledgment of the work of "native" informants, see Kapil Raj 2007. Referring to the work of Simon Schaffer (1994), Raj (2007, 216) notes how "at least until the mid-nineteenth century [Schaffer's period of focus], women, servants, children, the sick, and the insane were not considered to be reliable witnesses."

14. This also relates to critiques of the gendered treatment of victims of sexual violence. To paraphrase one well-known scenario first presented by Connie Borkenhagen (1975), imagine what would happen if a reported robbery were treated like a reported sexual assault. The victim might say, "A thief beat me up and stole my wallet," and the police would reply, "But what were you wearing? Why did you leave your home at all? Are sure you didn't just give your money away?"

15. In addition, this illustrates the importance of including Palestinians in decision making at an international level as well as developing international solidarities with other groups that face injustice. Solidarity is a prominent and sometimes-problematic part of antioccupation activism. On the challenges of forming ties of solidarity and advocating a more textured image for Palestinians in the United States, see Jordan 2002.

16. On Palestinian village histories, which are often accompanied by village atlases, see Davis 2011. These include the more recent atlas by Ali Hussein Abu Raya (2010), Salman Abu Sitta's (2007) richly detailed volume as well as the groundbreaking work of Khalidi (1992). For an example of an online interactive map of destroyed villages, see Miriyam Aouragh's (2011b) analysis of the Palestine Remembered project, as well

as the Nakba Archive (n.d.) and Arabic Cultural Association (n.d.). For examples of contemporary Palestinian atlases, see ARIJ 2009a; PalMap 2011. For examples of Israeli atlases, see Carta 2005; SOI 2009. On catastrophe and memory in Palestine, see Saloul 2012. On ideology and memory, see Ben-Ze'ev 2011. Prominent Palestinian scholars in memory studies include Kamal Abdel-Fattah, Sharif Kanaaneh, and Saleh Abd al-Jawwad. My thanks go to Miriyam Aouragh for pointing out several of these important sources and researchers.

17. In the same review, Brawer also criticizes Sami Hadawi and Walid Khalidi in tandem, as does Karsh (2000, 12n6). Brawer (1990) himself has investigated the negative impacts of border segregation on Palestinians, albeit in a context where the Israeli occupation was alleged to be preventing the "segregation" that might be introduced as a result of Palestinian self-rule.

18. Discursive power inequalities are inflected throughout legal, economic, and social imbalances. The legal imbalances are especially apparent. Even though all West Bank inhabitants are officially under the jurisdiction of military law, in practice this only affects Palestinians. Instead, the far more lenient and transparent civilian law is applied to Jewish Israeli residents (Military Court Watch 2013).

20. Even when power imbalances are less significant, different methods of observation can lead to varying results. This was seen in the controversy over whether Mount McKinley in Alaska was believed to have shrunk. The dispute stemmed in part from differences between the average height measurements taken by plane and point measurements attained by scaling McKinley's highest peak (Vergano 2013).

21. GPS devices find their own location through a not-unrelated process. A GPS device determines its own, unknown position through observations of four satellites with known orbits in space and time.

22. As we will see, this totalizing effect contrasts sharply with the post-1967 Israeli settlement project, which demonstrates how land and map cannot be extricated.

23. Fischer also thereby draws a structural parallel between observation, on the one hand, and symmetry as it is discussed in chapter 2, on the other.

24. Harding has come to focus increasingly on the academic literature on decolonization in an attempt to address these and related concerns.

25. This call encompasses differences between subjects as well as those between the scientist as an individual, on the one hand, and the data being produced, on the other. For as Niels van Doorn, Sally Wyatt, and Liesbet van Zoonen (2008) have pointed out, even if digital technologies do not sever the link between experience and embodiment, nonetheless the relationships between the body and text or data are not predetermined. Thus, these relationships are a fruitful subject for analysis in their own right.

26. Triangulation also parallels shifts in digital cartography from overhead surveys to data collection that indicates perspective, like Google Street View. This can be seen, for example, by comparing figure 3.2b with figure 4.5b. In figure 3.2b, data were collected in large consistent swathes, mainly from above. In figure 4.5b, data were collected at the street level, following roads and major trails on the ground.

27. In some ways, both figures 5.2a and 5.2b exceed even the access of international cartographers like those at the United Nations. These cartographers must travel in official vehicles and generally are not posted in one particular region for more than a few years, and so have less of the background knowledge as well as fluency in Arabic and Hebrew that many at ARIJ and Peace Now bring to their efforts (interview 17, UN cartographer). There are notable exceptions, such as Stefan Ziegler, the project manager of UNRWA's Barrier Monitoring Unit (interview 16, UN organization head).

28. Even the choice of which settlements to fly over and which sites to focus on are informed by Ofran's regular journeys within the settlements at ground level.

29. Although less frequent than arrests of Palestinian fieldworkers, in recent years Israeli activists have also been arrested for documenting settlement expansion (Zonszein 2013b).

30. To compare two maps of one Israeli area that exhibit varying levels of detail, see ARIJ 2013b; Peace Now 2012.

31. Similarly, Google Earth does not depict certain major military installations, and for this reason among others, its all-encompassing globe has inherent restrictions (Mogel and Bhagat 2007; Paglen 2009).

32. The launch of the Göktürk-3 satellite is planned for 2019, and it could have a major impact in terms of serving as a source for alternative forms of satellite imagery. Like its predecessors, Göktürk-1 and Göktürk-2, it has the potential to compete with the effective Israeli monopoly on the contents of public satellite images of Palestine and Israel—a monopoly whose restrictions are also enforced, per United States law, on companies located in the United States (Aleaziz 2011; Arianespace 2013; "Turkish Satellite" 2011).

33. For simplicity, I use the term *LULC maps* interchangeably with *land use maps*.

34. Overall there are substantial differences between Hadawi's and Amiran's work. Yet in this text, in ways that resonate with the writings of Amiran—whose negative characterization of Palestinian agriculture is discussed in chapter 2—and dominant development discourse at the time, Hadawi (1957, 17) refers to such methods as "still primitive." On British attempts to impose standardized cultivation in Palestine and Israel before 1948, see El-Eini 2006.

35. See also Walid Khalidi's (1993) reply to Benny Morris's (1992) review of Khalidi's (2004) *Before Their Diaspora*.

36. For an example of a case study of remote-sensing efforts carried out by scholars from Birzeit University, see Sharakas et al. 2007. Since roughly 2005, remote-sensing methods have increasingly gained traction in the West Bank as an alternative to field surveys, although the methods and their effects on land management are deserving of further study.

37. See also ARIJ's (n.d.-b) interactive settlement outpost map and ARIJ's (2009b) village profile map.

38. In 2011, a law was put to the Israeli Knesset that ultimately passed in a weaker form. Its goal was to monitor NGO funding sources, thereby increasing surveillance practices due to accusations that peace organizations are double agents for international governments. The move not only helped spark a series of mass protests in Israel but also was widely believed to be aimed at Peace Now (Oster 2011; Paraszczuk 2011; Rosenberg 2012; US Department of State 2010).

39. This is not to say that Israeli and Palestinian organizations can never work together, and Israeli NGOs are sometimes able to maintain contacts within Palestinian communities. These are the exceptions, though, and the produced landscapes in the region serve to discourage such efforts on multiple levels. My interlocutors regularly recounted the political and practical difficulties of attempts among organizations to support each other's efforts.

40. In addition, words like *cooperation* and *collaboration* have somber implications when used to refer to relationships with an occupying military. *Normalization* is also used much more broadly in the Middle East, where long campaigns were conducted against *tatbi'ah* (normalization) as a strategy to resist colonial attempts at the pacification and oppression of large populations. In this respect, Ofran's movements through the West Bank as well as those of specific other solidarity activists are exceptions to a broader politics of colonial movement on the part of settlers and the Israeli military. My thanks go to Miriyam Aouragh for pointing this out.

41. The ARIJ cartographers told me that they also spoke regularly with academics and observers in Gaza. However, due to the closure and permit restrictions, this took place via Skype and phone conversations rather than the more extensive in-person meetings they would have preferred. For a discussion of how this book is framed in part through the rigid closure of Gaza, see chapter 1.

42. For an insightful analysis of conceptions of movement and freedom in liberal political thought, and their relationship to subjectivities and modes of being, see Kotef 2015.

43. For a cogent critique of "transparent" forms of reflexivity, see Rose 1997.

44. This hesitancy to perpetuate domination likely also informs Ofran's reluctance to speak for the Palestinians. Additionally, her reaction is related to the debate over the *one-state* (as in one state of Palestine/Israel) versus the *two-state* (two separate

states, one Palestinian and one Israeli) solutions. This debate is also noted in chapter 1. To briefly summarize it once again: a two-state solution would see separate Palestinian and Israel states, whereas a one-state solution would, it is hoped, involve one binational state instead of a Jewish state and the current military occupation.

Peace Now unequivocally advocates a two-state solution with the aim of preserving a Jewish state, but separating this state from the West Bank and Gaza Strip with the aim of allowing them political sovereignty. Yet as peace talks continue to fail, and Israel continues to build settlements in the West Bank, the two-state solution increasingly seems hollow to many concerned with social justice. The literature on the one-state/two-state debate is enormous. For a prominent advocate of a one-state approach, see Abunimah 2007. For an argument that the one-state is already in effect through the occupation, see Azoulay and Ophir 2012. Noam Chomsky (2013) has argued that the one- versus two-state distinction is itself a fundamentally flawed distraction, but nonetheless it has important implications for sovereignty and the everyday lives of those living in the region.

45. Such alternative representations of space come through particularly effectively in graphic novels such as those by Oreet Ashery and Larissa Sansour (2009), Joe Sacco (2001, 2009), Rutu Modan (2008), and Etgar Keret and Assaf Hanuka (2005) as well as several on Jerusalem, including ones by Guy Delisle (2012) and Boaz Yakin and Nick Bertozzi (2013).

Chapter 6: The Geographic Production of Knowledge

1. The excerpt from Darwish's (2002) poem is reprinted with permission of Riad El-Rayyes Books. For an analysis of Palestinian and Jewish Israeli identity in Darwish's *Halat Hisar*, see Behar 2011.

2. For one of Darwish's own identity cards, see http://www.palestineremembered .com/Acre/al-Birwa/Picture88427.html (accessed October 21, 2015).

3. Geographic motifs abound in related literature and music. To give just two examples, see the song "At the Border" in Marcel Khalife's (1980) musical album of the same name, and the description of drawing maps on the ground from memory in Elias Khoury's (2005, 192) *Gate of the Sun*.

4. Indeed, the careful selection and refashioning of snippets of media are at the heart of both academic textual citations and sampling practices inspired by hip-hop, among other genres of art and music.

5. For a helpful discussion, see http://userpages.umbc.edu/~korenman/wmst/ masterstools.html (accessed December 12, 2015).

6. As I noted earlier, her analysis of the relationship between feminism and Islam is less helpful than critiques such as those by Chandra Talpade Mohanty, Ann Russo, and Lourdes Torres (1991) and Saba Mahmood (2005). Yet as I attempt to show, her

insights into the relationships between colonialism and patriarchy remain highly relevant.

7. There are many different ways to be international, and indeed many Palestinians and Israelis are themselves also internationals. The security measures carried out by the Israeli authorities overwhelmingly impact the ability of those academics who don't hold local passports or residence permits to carry out work there. Entrance might be rejected at any checkpoint, and it has been reported that Israeli airport authorities regularly search private e-mail and social media accounts in an attempt to gain access to information on those working to end the Israeli occupation (e.g., Friedman 2012; see also Hass 2011b).

8. Due in part to negative experiences with international aid, activists are beginning to use alternative sources of funds, including crowdsourcing through social media. In small and discrete cases, virtual mobility may in fact increase corporeal mobility. For example, Ezra Nawi successfully applied to the crowdfunding Web site Indiegogo to obtain financing for a new pickup truck to transport activists to protests; see http://www.indiegogo.com/projects/ezra-nawi-truck-campaign?c=home (accessed December 12, 2015). Yet crowdfunding is far from immune to power imbalances, since each backer judges a call for funds in part based on their own positioning. Indeed, crowdsourcing can lead to popularity contests of the type that democratic funding agencies ideally were intended to prevent.

9. This is by no means restricted to Egypt, where El Saadawi lives. In fact, the protections for academics in Europe and North America are being eroded through economic austerity measures, and critical scholars who focus on Palestine and Israel have long been under considerable political and economic pressure, exerted through a variety of practices from death threats to being denied or fired from research positions.

10. For a discussion of the Wall and its effects, see chapter 2.

References

Abbas, Mokarram, and Bas Van Heur. 2013. "Thinking Arab Women's Spatiality: The Case of 'Mutanazahat' in Nablus, Palestine." *Gender, Place, and Culture* 21 (10): 1–16.

Abdel Fattah, Awad. 2013. "Just Another Interrogation: My Encounter with the Shin Bet." *+972 Magazine,* April 4.

Abdelhamid, Ali. 2006. "Urban Development and Planning in the Occupied Palestinian Territories: Impacts on Urban Form." Paper presented at the Conference on Nordic and International Urban Morphology: Distinctive and Common Themes, Stockholm, September 3.

Abdullah, Abdullah S. 2005. *An Approach towards the Development of National Geographic Information Strategy in Palestine.* Ramallah: Birzeit University.

Abowd, Thomas. 2000. "The Moroccan Quarter: A History of the Present." *Jerusalem Quarterly* 7:6–16.

Abowd, Thomas. 2007. "Present and Absent: Historical Invention and the Politics of Place in Contemporary Jerusalem." In *Reapproaching Borders: New Perspectives on the Study of Israel–Palestine,* ed. Sandy Sufian and Mark Levine, 243–265. Lanham, MD: Rowman and Littlefield.

Abu El-Haj, Nadia. 2001. *Facts on the Ground: Archaeological Practice and Territorial Self-Fashioning in Israeli Society.* Chicago: University of Chicago Press.

Abu El-Haj, Nadia. 2002. "Producing (Arti) Facts: Archaeology and Power during the British Mandate of Palestine." *Israel Studies* 7 (2): 33–61.

Abujidi, Nurhan. 2011. "Surveillance and Spatial Flows in the Occupied Palestinian Territories." In *Surveillance and Control in Israel/Palestine: Population, Territory, and Power,* ed. Elia Zureik, David Lyon, and Yasmeen Abu-Laban, 313–334. London: Routledge.

Abujidi, Nurhan. 2014. *Urbicide in Palestine: Spaces of Oppression and Resilience.* London: Routledge.

Abu Kubi, Majed. 2003. "Detection and Mapping of the Land Use / Land Cover (LULC) Changes in the Jordan Valley Using Landsat Imageries." In *Environmental Monitoring in the South-Eastern Mediterranean Region Using RS/GIS Techniques*, ed. Ioannis Z. Gitas and Jesus San-Miguel-Ayanz, 69–84. Chania: Centre international de hautes études agronomiques méditerranéennes.

Abu-Libdeh, Hassan. 2001a. "An Urgent Appeal: The Israeli Military Forces Take over PCBS." Ramallah: PCBS.

Abu-Libdeh, Hassan. 2001b. Update No. 2, the Israeli Military Forces Withdraw from the PCBS Premises Leaving Behind Ruins, Stolen Property, Missing Files, Missing Computers, and Missing Datasets. Ramallah: PCBS.

Abu-Libdeh, Hassan. 2001c. Update No. 3, the Day After: PCBS Inspects Its Hardware for Harmful Devices. Ramallah: PCBS.

Abu-Libdeh, Hassan. 2001d. Update No. 4, War against the PCBS: The Days After, PCBS Response to IDF Claims. Ramallah: PCBS.

Abunimah, Ali. 2006. *One Country: A Bold Proposal to End the Israeli–Palestinian Impasse*. New York: Henry Holt.

Abu Raya, Ali Hussein. 2010. *Atlas Ma'alam Filistin Qibl 'Am 1948* [Atlas of the Landmarks of Palestine before 1948]. Sakhnin: Dar al-Ma'alam.

Abu Sitta, Salman. 2004. *The Atlas of Palestine, 1948*. London: Palestine Land Society.

Abu Sitta, Salman. 2007. *The Return Journey: A Guide to the Ethnically Cleansed and Present Palestinian Towns, Villages, and Holy Sites in English, Arabic, and Hebrew*. London: Palestine Land Society.

Abu-Zahra, Nadia, and Adah Kay, eds. 2012. *Unfree in Palestine: Registration, Documentation, and Movement Restriction*. London: Pluto Press.

Adalah. 2013. "Demolition and Eviction of Bedouin Citizens of Israel in the Naqab (Negev): The Prawer Plan." Adalah: The Legal Center for Arab Minority Rights in Israel, July.

Adas, Michael. 1989. *Machines as the Measure of Men: Science, Technology, and Ideologies of Western Dominance*. Ithaca, NY: Cornell University Press.

Adonis, Mahmoud Darwish, and Samih al-Qasim. 1995. *Victims of a Map: A Bilingual Anthology*. London: Saqi Books.

Agamben, Giorgio. 1998. *Homo Sacer: Sovereign Power and Bare Life*. Stanford, CA: Stanford University Press.

Agamben, Giorgio. 2005. *State of Exception*. Chicago: University of Chicago Press.

Ahlqvist, Ola. 2005. "Transformation of Geographic Information Using Crisp, Fuzzy, and Rough Semantics." In *Re-Presenting GIS*, ed. Peter Fisher and Donald J. Unwin, 99–112. Chichester, UK: John Wiley and Sons.

Ahmed, Sara. 2006. *Queer Phenomenology: Orientations, Objects, Others*. Durham, NC: Duke University Press.

Ahmed, Sara. 2010. "Orientations Matter." In *New Materialisms: Ontology, Agency, and Politics*, ed. Diana Coole and Samantha Frost, 234–257. Durham, NC: Duke University Press.

Ahmed, Sara. 2014. *Willful Subjects*. Durham, NC: Duke University Press.

Akerman, James R. 2006. *Cartographies of Travel and Navigation*. Chicago: University of Chicago Press.

Al-Araj, Sheerin. 2008. "Social Ties between the People of Al-Walaja Village at Home and Abroad." In *Crossing Borders, Shifting Boundaries: Palestinian Dilemmas*, ed. Sari Hanafi, 12–26. Cairo: American University in Cairo Press.

Alatout, Samer. 2009. "Bringing Abundance into Environmental Politics: Constructing a Zionist Network of Water Abundance, Immigration, and Colonization." *Social Studies of Science* 39 (3): 363–394.

Aleaziz, Hamed. 2011. "Why Google Earth Can't Show You Israel." *Mother Jones*, June, 10.

Alkhalili, Nura. 2012. "Contestation of Space." *Berkeley Planning Journal* 25 (1): 160–180.

Allen, Lori. 2008. "Getting by the Occupation: How Violence Became Normal during the Second Palestinian Intifada." *Cultural Anthropology* 23 (3): 453–487.

Allen, Lori. 2013. *The Rise and Fall of Human Rights: Cynicism and Politics in Occupied Palestine*. Stanford, CA: Stanford University Press.

Al-Mahfedi, Mohamed Hamoud Kassim. 2011. "Edward Said's 'Imaginative Geography' and Geopolitical Mapping: Knowledge/Power Constellation and Landscaping Palestine." *Criterion* 2 (3): 1–26.

Amiran, David. 1987. "Land Transformation in Israel." In *Land Transformation in Agriculture*, ed. Markley Gordon Wolman and F.G.A. Fournier, 291–317. New York: John Wiley and Sons.

Amit, Gish. 2008. "Ownerless Objects? The Story of the Books Palestinians Left Behind in 1948." *Jerusalem Quarterly* 33:7–20.

Amnesty International. 2003. *Israel and the Occupied Territories: Surviving under Siege: The Impact of Movement Restrictions on the Right to Work*. London: Amnesty International.

Anderson, Benedict. 2006. *Imagined Communities*. New York: Verso.

Antoon, Sinan. 2008. "Returning to the Wind: On Darwish's *La Ta'tadhir 'Amma Fa'alta*." In *Mahmoud Darwish, Exile's Poet: Critical Essays*, ed. Hala Khamis Nassar and Najat Rahman, 215–238. Northampton, MA: Olive Branch Press.

Aouragh, Miriyam. 2011a. "Confined Offline, Traversing Online: Palestinian Mobility through the Prism of the Internet." *Mobilities* 6 (3): 375–397.

Aouragh, Miriyam. 2011b. *Palestine Online: Transnationalism, Communications, and the Reinvention of Identity*. London: I. B.Tauris.

Aouragh, Miriyam. 2015. "Revolutionary Manoeuverings: Palestinian Activism between Cybercide and Cyber Intifada." In *Media and Political Contestation in the Contemporary Arab World: A Decade of Change*, ed. Lena Jayyusi and Anne Sofie Roald, 129–160. Houndmills, UK: Palgrave Macmillan.

Aouragh, Miriyam, and Paula Chakravartty. 2016. "Infrastructures of Empire: Towards a Critical Geopolitics of Media and Information Studies." *Media, Culture, and Society* 38 (4): 559–575.

Arabic Cultural Association. n.d. "The Witnessing Roots." http://www.roots48 -aca.org/media-eng/articles/270/The-Witnessing-Roots (accessed October 1, 2014).

Arianespace. 2013. "Arianespace to Launch Goturk-1 High-Resolution Observation Satellite." *Aerospace and Defense News*, June 19.

ARIJ (Applied Research Institute Jerusalem). 2000. *Attempt at Annexing the Modi'in Settlement Block*. Bethlehem: ARIJ.

ARIJ. 2004a. *Ecocide in [Taqu'a] Town*. Bethlehem: ARIJ.

ARIJ. 2004b. *Israeli Settlers Annex More Lands in [Taqu'a] Town*. Bethlehem: ARIJ.

ARIJ. 2006a. Israeli Military Orders Database. http://orders.arij.org (accessed October 1, 2016).

ARIJ. 2006b. *New Expansions in the Israeli West Bank Settlements*. Bethlehem: ARIJ.

ARIJ. 2008. "West Bank Land Use/Land Cover." Bethlehem: ARIJ.

ARIJ. 2009a. *A Geopolitical Atlas of the Occupied Palestinian Territory*. Bethlehem: ARIJ.

ARIJ. 2009b. "The Palestinian Locality Profiles." http://vprofile.arij.org (accessed October 1, 2016).

ARIJ. 2013a. *Israeli Authorities Ban Entry of ARIJ Intern*. Bethlehem: ARIJ.

ARIJ. 2013b. *The Location of Kfar Adumim Settlement and the Targeted Palestinian School*. Bethlehem: ARIJ.

ARIJ. n.d.-a "The Applied Research Institute Jerusalem (ARIJ) Deplores and Condemns the Death Threats against Hagit Ofran [...]." Bethlehem: ARIJ.

ARIJ. n.d.-b "Israeli Settlement Outposts in the West Bank." Bethlehem: ARIJ.

Ashery, Oreet, and Larissa Sansour. 2009. *The Novel of Nonel and Vovel*. Milan: Charta.

Atshan, Laila. 2011. "Keeping the Candle Lit." Bethlehem: TEDx Ramallah.

Azoulay, Ariella, and Adi Ophir. 2012. *The One-State Condition: Occupation and Democracy in Israel/Palestine*. Stanford, CA: Stanford University Press.

Bachi, Roberto. 1955. "The Graphical Representation of Geographical Series." *Bulletin of the International Statistical Institute* 34:455–470.

Bachi, Roberto. 1956. *Studies in Economic and Social Sciences. Scripta Hierosolymitana III*. Jerusalem: Magnes Press.

Bachi, Roberto. 1962a. "Some New Methods for the Graphical Representation of Statistical Data." *Bulletin of the International Statistical Institute* 40:343–361.

Bachi, Roberto. 1962b. "Standard Distance Measures and Related Methods for Spatial Analysis." In *Regional Science Association: Papers X*, 83–132. Zurich: Regional Science Association.

Bachi, Roberto. 1968. *Graphical Rational Patterns: A New Approach to Graphical Presentation of Statistics*. Jerusalem: Israel Universities Press.

Bachi, Roberto. 1974a. "The Jewish Population." In *Society*, 1–28. Jerusalem: Keter.

Bachi, Roberto. 1974b. *The Population of Israel*. Jerusalem: Institute of Contemporary Jewry.

Bachi, Roberto. 1975. "Graphical Methods for Presenting Statistical Data: Progress and Problems." In *Proceedings of the International Symposium on Computer-Assisted Cartography*, 74–98. Washington, DC: US Department of Commerce, Bureau of the Census.

Bachi, Roberto. 1980. "A Population Policy for Israel?" *Jewish Journal of Sociology* 22 (2): 163–179.

Bachi, Roberto. 1981. "Mapping the Characteristics of Distributions of Populations over Territories." *Bulletin of the International Statistical Institute* 49:1003–1026.

Bachi, Roberto. 1989. "Rational Maps and Parameters of Geographical-Statistical Data." *Espace, Populations, Sociétés* 7 (3): 337–348.

Bachi, Roberto. 1999. *New Methods of Geostatistical Analysis and Graphical Presentation: Distributions of Populations over Territories*. New York: Kluwer Academic.

Bachi, Roberto, Benjamin Gil, Helmut Mühsam, and Moshe Sicron. 1955. *Registration of Population, Part A: Towns, Villages, and Regions (Extract)*. Special Series No. 36(8) XI 1948. Jerusalem: Israel CBS.

Bahat, Kobi. 1997. "Census Mapping Using the GIS System." Paper presented at EUROSTAT and the Israeli Central Bureau of Statistics workshop, New Technologies for the 2000 Census Round: Sharing the Israeli Experience from the 1995 Census, Kibbutz Ma'ale Hachamisha, Israel, March 16.

Balibar, Etienne. 1990. "The Nation Form: History and Ideology." *RE:view* 13 (3): 329–361.

Bank, André, and Bas Van Heur. 2007. "Transnational Conflicts and the Politics of Scalar Networks: Evidence from Northern Africa." *Third World Quarterly* 28 (3): 593–612.

Barak, Aharon. 2005. *Mara'abe v. the Prime Minister of Israel*. The High Court of Justice of Israel, September 15.

Barak, Dafna. 1997. "Producing Maps and Data-Files for the Census and Post-Census Geographical Products." Paper presented at EUROSTAT and the Israeli Central Bureau of Statistics workshop, New Technologies for the 2000 Census Round: Sharing the Israeli Experience from the 1995 Census, Kibbutz Ma'ale Hachamisha, Israel, March 16.

Bar-Gal, Yoram. 2013. Prof. David Amiran (1910–2003). http://geo.haifa.ac.il/~bargal/history/english/amiran_english.htm (accessed October 1, 2016).

Barnard, Ryvka. 2013. "Seeing How the Natives Live: On the Pitfalls and Potential of Alternative Tourism." *Jadaliyya*, April 16.

Barnes, Barry, David Bloor, and John Henry. 1996. *Scientific Knowledge: A Sociological Analysis*. London: Athlone.

Barnes, Trevor J. 2001. "Lives Lived and Lives Told: Biographies of Geography's Quantitative Revolution." *Environment and Planning D: Society and Space* 19 (4): 409–429.

Baskin, Elisha, and Donna Nevel. 2013. "Thoughts about Our Role and Work as Jews Committed to Justice in Palestine." Mondoweiss, May 8.

Beaulieu, Anne, Sarah de Rijcke, and Bas Van Heur. 2013. "Authority and Expertise in New Sites of Knowledge Production." In *Virtual Knowledge: Experimenting in the Humanities and the Social Sciences*, ed. Paul Wouters, Anne Beaulieu, Andrea Scharnhorst, and Sally Wyatt, 25–56. Cambridge, MA: MIT Press.

Behar, Almog. 2011. "Mahmoud Darwish: Poetry's State of Siege." *Journal of Levantine Studies* 1 (1): 189–199.

Bekker, Vita. 2013. "Palestinian Activists Sue Israel for the Return of 6,000 Books." *National* (Abu Dhabi), May 19.

Bektas, Yakup. 2000. "The Sultan's Messenger: Cultural Constructions of Ottoman Telegraphy, 1847–1880." *Technology and Culture* 41 (4): 669–696.

Bender, Barbara, and Margot Winer, eds. 2001. *Contested Landscapes: Movement, Exile, and Place*. Oxford: Berg.

Benjamin, Walter. (1974) 2001. "On the Concept of History." *Gesammelten Schriften*, I:2. Trans. Dennis Redmond. Suhrkamp Verlag. Frankfurt am Main.

Ben-Moshe, Eliahu. 1997a. "Integration of a National GIS Project within the Planning and Implementation of a Population Census." Paper presented at EUROSTAT and the Israeli Central Bureau of Statistics workshop, New Technologies for the 2000 Census Round: Sharing the Israeli Experience from the 1995 Census, Kibbutz Ma'ale Hachamisha, Israel, March 16.

Ben-Moshe, Eliahu. 1997b. "The Israeli 1995 Census–GIS Integration Project." Paper presented at the Work Session on Geographical Information Systems of the Statistical Commission and Economic Commission for Europe Conference of European Statisticians, Brighton, UK, September 22.

Ben-Moshe, Eliahu. 1997c. "Planning a Census by the End of the Nineties." Paper presented at EUROSTAT and the Israeli Central Bureau of Statistics workshop, New Technologies for the 2000 Census Round: Sharing the Israeli Experience from the 1995 Census, Kibbutz Ma'ale Hachamisha, Israel, March 16.

Benvenisti, Meron. 1984. "US Government Funded Projects in the West Bank and Gaza (1977–1983)." Working Paper 13. Jerusalem: West Bank Data Base Project.

Benvenisti, Meron. 1999. "Bikini on Jerusalem's Beach." *Ha'aretz*, July 29.

Benvenisti, Meron. 2000. *Sacred Landscape: The Buried History of the Holy Land since 1948*. Berkeley: University of California Press.

Benvenisti, Meron, and Shlomo Khayat. 1988. *The West Bank and Gaza Atlas*. Jerusalem: West Bank Data Base Project.

Ben-Ze'ev, Efrat. 2011. *Remembering Palestine in 1948: Beyond National Narratives*. Cambridge: Cambridge University Press.

Benziman, Uzi. 2006. "Moving on to the Next Scandal …" *Ha'aretz*, December 20.

Bhabha, Homi K. 2004. *The Location of Culture*. London: Routledge.

Bhungalia, Lisa. 2009. "A Liminal Territory: Gaza, Executive Discretion, and Sanctions Turned Humanitarian." *GeoJournal* 75 (4): 347–357.

Bhungalia, Lisa. 2012. "Im/Mobilities in a 'Hostile Territory': Managing the Red Line." *Geopolitics* 17 (2): 256–275.

Bier, Jess. 2017. "Palestinian State Maps and Imperial Technologies of Staying Put." *Public Culture* 29 (181): 53–78.

Bijker, Wiebe. 1995. *Of Bicycles, Bakelites, and Bulbs: Toward a Theory of Sociotechnical Change.* Cambridge, MA: MIT Press.

Bijker, Wiebe, Roland Bal, and Ruud Hendriks. 2009. *The Paradox of Scientific Authority: The Role of Scientific Advice in Democracies.* Cambridge, MA: MIT Press.

Bishara, Amahl. 2003. "House and Homeland: Examining Sentiments about and Claims to Jerusalem and Its Houses." *Social Text* 21 (2): 141–162.

Blake, Gerald. 1995. "The Depiction of International Boundaries on Topographic Maps." *International Boundaries Research Unit Boundary and Security Bulletin*, 44–50.

Bland, Sally. 2004. "Sami Hadawi: The Scholar Who Couldn't Go Home." *Middle East Window and Jordan Times*, May 17.

Bloor, David. 1976. *Knowledge and Social Imagery.* London: Routledge and Kegan Paul.

Blum, Olivia. 1997. "Logical Structure and Guiding Principles." Paper presented at EUROSTAT and the Israeli Central Bureau of Statistics workshop, New Technologies for the 2000 Census Round: Sharing the Israeli Experience from the 1995 Census, Kibbutz Ma'ale Hachamisha, Israel, March 16.

Boersma, Sanne, and Willem Schinkel. 2015. "Imagining Society: Logics of Visualization in Images of Immigrant Integration." *Environment and PlanningD: Society and Space* 33 (6): 1043–1062.

Borkenhagen, Connie K. 1975. "The Legal Bias against Rape Victims." *American Bar Association Journal* (April): 464.

Bossard, Michel, Jan Feranec, and Jan Otahel. 2000. *Corine Land Cover Technical Guide: Addendum 2000.* Copenhagen: European Environment Agency.

Bouillon, Markus E. 2004. *The Peace Business: Money and Power in the Palestine–Israel Conflict.* London: I. B. Tauris.

Bowler, Peter J., and Iwan Rhys Morus. 2005. *Making Modern Science: A Historical Survey.* Chicago: University of Chicago Press.

Braidotti, Rosi. 2013. *Nomadic Subjects: Embodiment and Sexual Difference in Contemporary Feminist Theory.* New York: Columbia University Press.

Braverman, Irus. 2009. "Planting the Promised Landscape: Zionism, Nature, and Resistance in Israel/Palestine." *Natural Resources Journal* 49 (2): 317–361.

Brawer, Moshe. 1990. "The 'Green Line': Functions and Impacts of an Israeli—Arab Superimposed Boundary." In *International Boundaries and Boundary Conflict Resolution*, ed. Carl Grundy-Warr, 63–74. Durham, UK: Boundaries Research Press.

Brawer, Moshe. 1994. "All That Remains?" *Israel Affairs* 1 (2): 334–345.

Brawer, Moshe. 2008. "The Image of Israel's Geographical Transformation, in Honor of Israel Prize Recipient, Prof. Elisha Efrat." *Israel Studies* 13 (1): 152–159.

Bronner, Ethan. 2008. "3 Fulbright Winners in Gaza Again Told They Can't Travel." *New York Times*, August 5.

Brown, Wendy. 2010. *Walled States, Waning Sovereignty*. New York: Zone Books.

Bshara, Khaldun A. M. 2012. "Space and Memory: The Poetics and Politics of Home in the Palestinian Diaspora." PhD diss., University of California at Irvine.

B'Tselem. 2004. *Forbidden Roads: Israel's Discriminatory Road Regime in the West Bank.* Jerusalem: B'Tselem.

B'Tselem and Eyal Weizman. 2002. *Jewish Settlements in the West Bank: Built-up Areas and Land Reserves.* Jerusalem: B'Tselem.

Buckley, Craig, and Mark Wasiuta, eds. 2013. *Collecting Architecture Territories.* New York: GSAPP / Deste Foundation for Contemporary Art.

Bunge, William. 1966. "Gerrymandering, Geography, and Grouping." *Geographical Review* 56 (2): 256–263.

Butler, Judith. 2015. *Notes toward a Performative Theory of Assembly.* Cambridge, MA: Harvard University Press.

Büttner, George. Jan Feranec, Gabriel Jaffrain, Chris Steenmans, Adriana Gheorghe, and Vanda Lima. 2002. *Corine Land Cover Update 2000: Technical Guidelines.* Copenhagen: European Environment Agency.

Callon, Michael, and Bruno Latour. 1992. "Don't Throw the Baby Out with the Bath School! A Reply to Colins and Yearly." In *Science as Practice and Culture*, ed. Andrew Pickering. Chicago: University of Chicago Press.

Calvo, Rinat. 1997. "Redistricting Enumeration Areas and Defining the Organizational Structure." Paper presented at the EUROSTAT and the Israeli Central Bureau of Statistics workshop, New Technologies for the 2000 Census Round: Sharing the Israeli Experience from the 1995 Census, Kibbutz Ma'ale Hachamisha, Israel, March 16.

Cameron, Angus. 2011. "Ground Zero: The Semiotics of the Boundary Line." *Social Semiotics* 21 (3): 417–434.

Carmon, Naomi. 1999. "Three Generations of Urban Renewal Policies: Analysis and Policy Implications." *Geoforum* 30 (2): 145–158.

Carta. 2005. *Super Atlas Carta Eretz Israel* [Carta's Super Atlas of the Land of Israel]. Jerusalem: Carta.

CBS (Central Bureau of Statistics). 1963. "Average Number of Persons in Household by Natural Region." In *The Settlements of Israel, Part II: Sex, Age, Residence, and Population Group: Data from Stage A of the Census*. Vol. 11: n.p., *1961 Population and Housing Census*. Jerusalem: Israeli Central 369 CBS.

CBS. 1967. *West Bank of the Jordan, Gaza Strip, and Northern Sinai, Golan Heights: Data from Full Enumeration*. Census of Population, 1967: Publication No. 1. Jerusalem: CBS.

CBS. 1969a. Distribution of the Jewish and Non-Jewish Population in 1968 by Locality. Jerusalem: CBS. GL-3556/16. 94.0/1-226. 02-105-06-04-09. Israel State Archives.

CBS. 1969b. 1972 Census of Population and Housing: Methodology. Jerusalem: CBS. 7-731-1322-1972-0011/8, 29/10/1968-3/10/1969. Israel State Archives.

CBS. 1985. *Population and Households, Socio-Economic Characteristics from the Sample Enumeration*. Vol. 5, *1983 Census of Population and Housing*. Jerusalem: CBS.

CBS. 2000a. Census 1995. Jerusalem: CBS GIS Unit. http://gis.cbs.gov.il/website/eng/viewer.htm (accessed October 1, 2016).

CBS. 2000b. *Characterization and Classification of Geographical Units by the Socio-Economic Level of the Population*. Vol. 13, *1995 Census of Population and Housing*. Jerusalem: CBS.

CBS. 2012. "Monthly Bulletin of Statistics of Israel, Population Table B-1: Population by Population Group, Thousands." December.

Chadwick, Alex. 2006. "Israeli Group: Leaked Maps Counter Government." National Public Radio News, November 21.

Chomsky, Noam. 2012. "My Visit to Gaza, the World's Largest Open-Air Prison." *Truthout*, November 9.

Chomsky, Noam. 2013. "Chomsky: Don't Be Distracted by One-State/Two-State 'Debate' on Israel—Something Much More Nefarious Is Going On." *AlterNet*, November 4.

Clark, Nigel, Doreen B. Massey, and Philip Sarre, eds. 2008. *Material Geographies: A World in the Making*. London: Sage.

Claudet, Sophie. 2001. "Israeli Army Raid Sets Dangerous New Precedent." *Daily Star* (Beirut), December 10.

Clifford, James. 2001. "Indigenous Articulations." *Contemporary Pacific* 13 (2): 468–490.

Clifford, James. 2003. *On the Edges of Anthropology: Interviews*. Chicago: Prickly Paradigm Press.

Cohen, Shaul. 1993. *The Politics of Planting: Israeli–Palestinian Competition for Control of Land in the Jerusalem Periphery*. Chicago: University of Chicago Press.

Cohen, Shaul. 1994. "Greenbelts in London and Jerusalem." *Geographical Review* 84 (1): 74–89.

Cohen, Shaul. 2000. "An Absence of Place: Expectation and Realization in the West Bank." In *Cultural Encounters with the Environment: Emerging and Evolving Geographic Themes*, ed. Alexander B. Murphy, Douglas L. Johnson, and Viola Haarman, 283–303. Lanham, MD: Rowman and Littlefield.

Cohen, Shaul. 2002. "As a City Besieged: Place, Zionism, and the Deforestation of Jerusalem." *Environment and Planning D: Society and Space* 20 (2): 209–230.

Coignet, Gildas. 2009. "Espaces Publics et Identité Nationale, de la Capitale Arabe Moderne à la Métropole Mondialisée" [Public Spaces and National Identity: From the Modern Arab Capital to the Globalized Metropolis]. *Les Cahiers d'EMAM: Études sur le Monde Arabe et la Méditerranée* 18:123–126.

Coogan, Mike. 2013. Senator Boxer's Farfetched Defense of the Visa Waiver Exemption for Israel. Mondoweiss, May 10.

Coon, Anthony G. 1990. "Development Plans in the West Bank." *GeoJournal* 21 (4): 363–373.

Cosgrove, Denis. 2007. "Epistemology, Geography, and Cartography: Matthew Edney on Brian Harley's Cartographic Theories." *Annals of the Association of American Geographers* 97 (1): 202–209.

Cosgrove, Denis. 2008. *Geography and Vision: Seeing, Imagining, and Representing the World* . London: I. B. Tauris.

Courbage, Youssef. 1999. "Reshuffling the Demographic Cards in Israel/Palestine." *Journal of Palestine Studies* 28 (4): 21–39.

Coutard, Olivier, and Simon Guy. 2007. "STS and the City Politics and Practices of Hope." *Science, Technology, and Human Values* 32 (6): 713–734.

Craib, Raymond B. 2009. "Relocating Cartography." *Postcolonial Studies* 12 (4): 481–490.

Crampton, Jeremy W. 2001. "Maps as Social Constructions: Power, Communication, and Visualization." *Progress in Human Geography* 25:235–252.

Crampton, Jeremy W. 2003. "Cartographic Rationality and the Politics of Geosurveillance and Security." *Cartography and Geographic Information Science* 30 (2): 135–148.

Crampton, Jeremy W. 2010. *Mapping: A Critical Introduction to Cartography and GIS*. Malden, MA: Wiley-Blackwell.

Crampton, Jeremy W., and Stuart Elden, eds. 2007. *Space, Knowledge, and Power: Foucault and Geography*. Aldershot, UK: Ashgate.

Crampton, Jeremy W., and John Krygier. 2006. "An Introduction to Critical Cartography." *Acme* 4 (1): 11–33.

Cranor, John D., Gary L. Crawley, and Raymond H. Scheele. 1989. "The Anatomy of a Gerrymander." *American Journal of Political Science* 33 (1): 222–239.

Cresswell, Tim. 1999. "Embodiment, Power, and the Politics of Mobility: The Case of Female Tramps and Hobos." *Transactions of the Institute of British Geographers* 24 (2): 175–192.

Cresswell, Tim. 2011. "Mobilities I: Catching Up." *Progress in Human Geography* 35 (4): 550–558.

Culcasi, Karen. 2008. "Cartographic Constructions of the Middle East." PhD diss., Syracuse University.

Culcasi, Karen. 2010. "Constructing and Naturalizing the Middle East." *Geographical Review* 100 (4): 583–597.

Culcasi, Karen. 2012. "Mapping the Middle East from Within: (Counter-) Cartographies of an Imperialist Construction." *Antipode* 44 (4): 1099–1118.

Curry, Michael R. 1998. *Digital Places: Living with Geographic Information Technologies*. London: Routledge.

Daoud, Ribhi M. 2000. "An Evaluation of the Role of the Palestine Economic Council for Development and Reconstruction (PECDAR) in Palestine's Infrastructure Development since the 1993 Oslo Accords." PhD diss., Walden University.

Darwish, Mahmoud. 1993. *Ara Ma Urid* [I See What I Want]. Beirut: Dar al-Awdat.

Darwish, Mahmoud. 2002. *Halat Hisar* [State of Siege]. Beirut: Riad El-Rayyes Books.

Darwish, Mahmoud. 2003. *Unfortunately, It Was Paradise*. Trans. Munir Akash, Carolyn Forché, Sinan Antoon, and Amira El-Zein. Berkeley: University of California Press.

Darwish, Mahmoud. 2004. *La Ta'tadhir 'Amma Fa'alta* [Do Not Apologize for What You Have Done]. Beirut: Riad El-Rayyes Books.

Darwish, Mahmoud. 2005. *Al-Diwan: Al-A'amal Al-Ula* [Volumes: The Early Works]. Beirut: Riad El-Rayyes Books.

Darwish, Mahmoud. 2007. *The Butterfly's Burden: Poems*. Trans. F. Joudah. Port Townsend, WA: Copper Canyon Press.

Daston, Lorraine J., and Peter Galison. 2007. *Objectivity*. New York: Zone Books.

Dattel, Lior. 2013. "In the World of Charity, Israel Is Still Receiving a Lot More Than It Gives Back." *Ha'aretz*, May 23.

Davis, Angela Y. 1971. "Lessons from Attica to Soledad." *New York Times*, October 8.

Davis, Rochelle. 2011. *Palestinian Village Histories: Geographies of the Displaced.* Stanford, CA: Stanford University Press.

Davis, Uri, Antonia E. L. Maks, and John Richardson. 1980. "Israel's Water Policies." *Journal of Palestine Studies* 9 (2): 3–31.

de Vet, Annelys, ed. 2007. *Subjective Atlas of Palestine.* Rotterdam: 010 Publishers.

Delisle, Guy. 2012. *Jerusalem: Chronicles of the Holy City.* Montreal: Drawn and Quarterly.

Denzin, Norman K. 1970. *The Research Act in Sociology: A Theoretical Introduction to Sociological Methods.* London: Butterworths.

Denzin, Norman K. 1977. *The Research Act: A Theoretical Introduction to Sociological Methods.* New York: McGraw-Hill.

Derrida, Jacques. 1981. *Positions.* Trans. Alan Bass. Chicago: University of Chicago Press.

Dodge, Martin, Rob Kitchin, and Chris Perkins, eds. 2009. *Rethinking Maps: New Frontiers in Cartographic Theory.* New York: Routledge.

Doerr, Arthur H., Jerome F. Coling, and William S. Kerr. 1970. "Agricultural Evolution in Israel in the Two Decades since Independence." *Middle East Journal* 24 (3): 319–337.

Domosh, Mona. 1996. *Invented Cities: The Creation of Landscape in Nineteenth-Century New York and Boston.* New Haven, CT: Yale University Press.

Dorrian, Mark, and Gillian Rose, eds. 2003. *Deterritorialisations … Revisioning: Landscapes and Politics.* London: Black Dog.

Duncan, Sally L. 2006. "Mapping Whose Reality? Geographic Information Systems (GIS) and 'Wild Science.'" *Public Understanding of Science* 15 (4): 411–434.

Dunne, Christine E., Peter J. Atkins, Michael J. Blakemore, and Janet G. Townsend. 1999. "Teaching Geographical Information Handling Skills for Lower-Income Countries." *Transactions in GIS* 3 (4): 319–332.

Dweib, Riham Hassan. 2011. "Environmental Impacts in Tuklarem Governorate." MA thesis, Bir Zeit University.

Edgerton, David. 2007. "Creole Technologies and Global Histories: Rethinking How Things Travel in Space and Time." *Host: Journal of the History of Science and Technology* 1:75–112.

Edney, Matthew H. 1997. *Mapping an Empire: The Geographical Construction of British India, 1765–1843*. Chicago: University of Chicago Press.

Edney, Matthew H. 2005. "Putting 'Cartography' into the History of Cartography: Arthur H. Robinson, David Woodward, and the Creation of a Discipline." *Cartographic Perspectives* 51:14–29.

Edney, Matthew H. 2009. "The Irony of Imperial Mapping." In *The Imperial Map: Cartography and the Mastery of Empire*, ed. James R. Akerman, 11–45. Chicago: University of Chicago Press.

Efrat, Elisha. 1994. "Israel's Planned New 'Crossing Highway.'" *Journal of Transport Geography* 2 (4): 274–277.

El-Eini, Roza. 2006. *Mandated Landscape: British Imperial Rule in Palestine, 1929–1948*. New York: Routledge.

Elia, Nada. 2004. "Epistemic Violence, Smear-Campaigns, and Hit-Lists: Disappearing the Palestinians." In *Cultural Shaping of Violence: Victimization, Escalation, Response*, ed. Myrdene Anderson, 187–192. West Lafayette, IN: Purdue University Press.

Elia, Nada. 2005. "The Burden of Representation: When Palestinians Speak Out." *Journal of Middle East Studies* 5:58–70.

El Saadawi, Nawal. 1997. *The Nawal El Saadawi Reader*. London: Zed Books.

Elwood, Sarah. 2009. "Geographic Information Science: New Geovisualization Technologies, Emerging Questions, and Linkages with GIScience Research." *Progress in Human Geography* 33 (2): 256–263.

Engineers without Borders Palestine. n.d. "Free-Map Palestine—Bethlehem (and JUMP-START-INT)." http://www.ewb-palestine.org/Projects-2.html (accessed October 1, 2016).

Esack, Farid. 2009. "An Injury to One ..." *The Electronic Intifada*, May 10.

Escobar, Arturo. 2011. *Encountering Development: The Making and Unmaking of the Third World*. Princeton, NJ: Princeton University Press.

Essex, Jamey. 2008. "Deservedness, Development, and the State: Geographic Categorization in the US Agency for International Development's Foreign Assistance Framework." *Geoforum* 39 (4): 1625–1636.

Etkes, Dror, and Hagit Ofran. 2006. *Breaking the Law in the West Bank—One Violation Leads to Another: Israeli Settlement Building on Private Palestinian Property*. Jerusalem: Peace Now.

Ettinger, Yair. 2007. "Radical Black Panther Group Buries Sa'adia Marciano, 'The Face of Protest.'" *Ha'aretz*, December 23.

Eviction in East Jerusalem. 2009. "Inside Story." *Al Jazeera English*, February 24.

Fabian, Johannes. 2002. *Time and the Other: How Anthropology Makes Its Object.* New York: Columbia University Press.

Falah, Ghazi. 1985. "How Israel Controls the Bedouin in Israel." *Journal of Palestine Studies* 14 (2): 35–51.

Falah, Ghazi. 1989. "Israeli State Policy toward Bedouin Sedentarization in the Negev." *Journal of Palestine Studies* 18 (2): 71–91.

Fanon, Frantz. 2008. *Black Skin, White Masks.* New York: Grove Press.

Farah, Joseph. 2002. "Palestinian People Do Not Exist." *World Net Daily*, July 11.

Farman, Jason. 2010. "Mapping the Digital Empire: Google Earth and the Process of Postmodern Cartography." *New Media and Society* 12 (6): 869–888.

Feldman, Ilana. 2008. *Governing Gaza.* Durham, NC: Duke University Press.

Ferguson, James. 2012. "Structures of Responsibility." *Ethnography* 13 (4): 558–562.

Fields, Gary. 2010. "Enclosure: Palestinian Landscape in a 'Not-Too-Distant Mirror.'" *Journal of Historical Sociology* 23 (2): 216–250.

Fields, Gary. 2012. "'This Is Our Land': Collective Violence, Property Law, and Imagining the Geography of Palestine." *Journal of Cultural Geography* 29 (3): 267–291.

Fischbach, Michael R. 2003. *Records of Dispossession: Palestinian Refugee Property and the Arab–Israeli Conflict.* New York: Columbia University Press.

Fischbach, Michael R. 2011. "British and Zionist Data Gathering on Palestinian Arab Landownership and Population during the Mandate." In *Surveillance and Control in Israel/Palestine: Population, Territory, and Power*, ed. Elia Zureik, David Lyon, and Yasmeen Abu-Laban, 297–312. London: Routledge.

Fischer, Michael M. J. 2006. "Changing Palestine—Israel Ecologies: Narratives of Water, Land, Conflict, and Political Economy, Then and Now, and Life to Come." *Cultural Politics* 2 (2): 159–192.

Fischhendler, Itay. 2008. "When Ambiguity in Treaty Design Becomes Destructive: A Study of Transboundary Water." *Global Environmental Politics* 8 (1): 111–136.

Fisher, Peter, Alexis Comber, and Richard Wadsworth. 2005. "Land Use and Land Cover: Contradiction or Complement." In *Re-Presenting GIS*, ed. Peter Fisher and David J. Unwin, 85–98. Chichester, UK: John Wiley and Sons.

Fisher, Peter, and Donald J. Unwin, eds. 2005. *Re-Presenting GIS.* Chichester, UK: John Wiley and Sons.

Forte, Tania. 2003. "Sifting People, Sorting Papers: Academic Practice and the Notion of State Security in Israel." *Comparative Studies of South Asia, Africa, and the Middle East* 23 (1): 215–223.

Foster, Zachary J. 2013. Ottoman and Arab Maps of Palestine, 1880s–1910s. *Afternoon Map*, July 30.

Foucault, Michel. 1986. "Of Other Spaces." *Diacritics* 16:22–27.

Foucault, Michel. 1998. *The History of Sexuality, Volume 1: The Will to Knowledge.* London: Penguin.

Friedman, Lara. 2012. *What You Need to Know about E-1.* Washington, DC: Americans for Peace Now.

Galeano, Eduardo. 1997. *Open Veins of Latin America: Five Centuries of the Pillage of a Continent.* New York: Monthly Review Press.

Galeano, Eduardo. 2000. *Upside Down: A Primer for the Looking-Glass World.* New York: Metropolitan Books.

Garb, Yaakov. 2004. "Constructing the Trans-Israel Highway's Inevitability." *Israel Studies* 9 (2): 180–217.

Gasteyer, Stephen P., and Cornelia Butler Flora. 2000. "Modernizing the Savage: Colonization and Perceptions of Landscape and Lifescape." *Sociologia Ruralis* 40 (1): 128–149.

Gavish, Dov. 1996. "Foreign Intelligence Maps: Offshoots of the 1:100,000 Topographic Map of Israel." *Imago Mundi* 48:174–184.

Gavish, Dov. 2005. *A Survey of Palestine under the British Mandate, 1920–1948.* New York: Routledge.

Gavish, Dov. 2006. "Barrel in the Courtyard of the Survey of Israel." Paper presented to the Working Group on the History of Colonial Cartography in the 19th and 20th Centuries of the International Cartographic Association at the International Symposium on "Old Worlds–New Worlds": The History of Colonial Cartography 1750–1950, Utrecht, August 21–23.

Gavish, Dov, and Ron Adler. 1999. *50 Years of Mapping Israel, 1948–1999.* Tel Aviv: SOI.

Gavish, Dov, and Avishai Ben-Porath. 2003. "Mapot Emek ha-Hulah: Saman Derek bi-Mipuwi Otimani ba-Eretz Israel" [The Hulah Valley Map of 1887: A Landmark in Ottoman Cartography in Palestine]. *Cathedra* (109): 131–138.

Ghussein, Walaa, and Annie Robbins. 2013. "Two Friends Meet for 5 Minutes in Jerusalem." Mondoweiss, May 19.

Gibson-Graham, J. K. 2002. "Beyond Global vs. Local: Economic Politics Outside the Binary Frame." In *Geographies of Power: Placing Scale*, ed. Andrew Herod and Melissa W. Wright, 25–60. Malden, MA: Blackwell.

Gieryn, Thomas F. 1999. *Cultural Boundaries of Science: Credibility on the Line*. Chicago: University of Chicago Press.

Gieryn, Thomas F. 2000. "A Space for Place in Sociology." *Annual Review of Sociology* 26:463–496.

Gieryn, Thomas F. 2006. "City as Truth-Spot: Laboratories and Field-Sites in Urban Studies." *Social Studies of Science* 36 (1): 5–38.

GIS History Project. 1997. "The GBF/DIME Case Study." http://www.ncgia.buffalo.edu/gishist/DIME.html (accessed September 21, 2015).

Gisha. 2012. "What Is the 'Separation Policy'?" Tel Aviv: Gisha Legal Center for Freedom of Movement.

Godlewska, Anne. 1999. *Geography Unbound: French Geographic Science from Cassini to Humboldt*. Chicago: University of Chicago Press.

Godlewska, Anne, and Neil Smith, eds. 1994. *Geography and Empire*. Oxford: Blackwell.

Goering, Kurt. 1979. "Israel and the Bedouin of the Negev." *Journal of Palestine Studies* 9 (1): 3–20.

Goffman, Erving. 1989. "On Fieldwork." *Journal of Contemporary Ethnography* 18 (2): 123–132.

Golan, Tal. 2004. "Introduction." *Israel Studies* 9 (2): iv–viii.

"Golda Meir Scorns Soviets." 1969. *Washington Post*, June 16.

Goldshleger, Naftaly, Irit Amit-Cohen, and Maxim Shoshany. 2006. "A Step Ahead of Time: Design, Allocation, and Preservation of Private Open Space in the 1920s— the Case of a Garden Suburb in Israel." *GeoJournal* 67 (1): 57–69.

Gordon, Neve. 2011. "Israel's Emergence as a Homeland Security Capital." In *Surveillance and Control in Israel/Palestine: Population, Territory, and Power*, ed. Elia Zureik, David Lyon and Yasmeen Abu-Laban, 153–170. London: Routledge.

Gorenberg, Gershom. 2012. "Draw the Line: How Israel Erases Itself." *Daily Beast*, March 26.

Gorney, Edna. 2007. "(Un)Natural Selection: The Drainage of the Hula Wetlands, an Ecofeminist Reading." *International Feminist Journal of Politics* 9 (4): 465–474.

Government of the United Kingdom. 1924. *Railway Clearing House, Coaching Arrangements Book: Passenger Traffic Section, Embracing the General Regulations Relating to Fares, Tickets, Luggage, Etc.* London: Jas. Truscott and Son.

Graham, Stephen. 2011. "Laboratories of War: Surveillance and US–Israeli Collaboration in War and Security." In *Surveillance and Control in Israel/Palestine: Population, Territory, and Power,* ed. Elia Zureik, David Lyon, and Yasmeen Abu-Laban, 133–152. London: Routledge.

Gratien, Chris. 2013. "Mapping Minorities in Syria." *Afternoon Map,* August 29.

Gregory, Derek. 1994. *Geographical Imaginations.* Cambridge, MA: Blackwell.

Guarnieri, Mya. 2013. "Palestinian University Shuts Down in Wake of Violence, Teacher Strikes." *+972 Magazine,* March 16.

Gurvitz, Yossi. 2011. "Palestinians Handcuffed, Detained for Picking Wildflowers." *+972 Magazine,* March 8.

Hadawi, Sami. 1957. *Land Ownership in Palestine.* New York: Palestine Arab Refugee Office.

Hadawi, Sami. 1970. *Village Statistics 1945: A Classification of Land and Area Ownership in Palestine.* Beirut: PLO Research Center.

Haffner, Jeanne. 2013. *The View from Above: The Science of Social Space.* Cambridge, MA: MIT Press.

Halabi, Usama. 2011. "Legal Analysis and Critique of Some Surveillance Methods Used by Israel." In *Surveillance and Control in Israel/Palestine: Population, Territory, and Power,* ed. Elia Zureik, David Lyon, and Yasmeen Abu-Laban, 199–218. London: Routledge.

Halffman, Willem. 2003. "Boundaries of Regulatory Science: Eco/toxicology and Aquatic Hazards of Chemicals in the US, England, and the Netherlands, 1970–1995." PhD diss., University of Amsterdam.

Halper, Jeff. 2006. "A Strategy within a Non-Strategy: Sumud, Resistance, Attrition, and Advocacy." *Journal of Palestine Studies* 35 (3): 45–51.

Halpern, Orit. 2015. *Beautiful Data: A History of Vision and Reason since 1945.* Durham, NC: Duke University Press.

Hamdi, Tahrir. 2011. "Bearing Witness in Palestinian Resistance Literature." *Race and Class* 52 (3): 21–42.

Hammerman, Ilana. 2011. "Illegal in Their Own Country." *Ha'aretz,* September 9.

Hanafi, Sari. 2005. *The Emergence of a Palestinian Globalized Elite: Donors, International Organizations, and Local NGOs.* Jerusalem: Institute of Jerusalem Studies.

Hanafi, Sari. 2009. "Spacio-cide: Colonial Politics, Invisibility, and Rezoning in Palestinian Territory." *Contemporary Arab Affairs* 2 (1): 106–121.

Hanafi, Sari, and Linda Tabar. 2003. "The Intifada and the Aid Industry: The Impact of the New Liberal Agenda on the Palestinian NGOs." *Comparative Studies of South Asia, Africa, and the Middle East* 23 (1–2): 205–214.

Handel, Ariel. 2009. "Where, Where to, and When in the Occupied Territories: An Introduction to Geography of Disaster." In *The Power of Inclusive Exclusion: Anatomy of Israeli Rule in the Occupied Palestinian Territories*, ed. Adi Ophir, Michal Givoni and Sari Hanafi, 179–222. New York: Zone Books.

Handel, Ariel. 2011. "Exclusionary Surveillance and Spatial Uncertainty in the Occupied Palestinian Territories." In *Surveillance and Control in Israel/Palestine: Population, Territory, and Power*, ed. Elia Zureik, David Lyon, and Yasmeen Abu-Laban, 199–218. London: Routledge.

Hannah, Matthew G. 2001. "Sampling and the Politics of Representation in US Census 2000." *Environment and Planning D: Society and Space* 19 (5): 515–534.

Hannah, Matthew G. 2009. "Calculable Territory and the West German Census Boycott Movements of the 1980s." *Political Geography* 28 (1): 66–75.

Hannam, Kevin, Mimi Sheller, and John Urry. 2006. "Editorial: Mobilities, Immobilities, and Moorings." *Mobilities* 1 (1): 1–22.

Haraway, Donna. 1988. "Situated Knowledges: The Science Question in Feminism and the Privilege of Partial Perspective." *Feminist Studies* 14:575–599.

Haraway, Donna. 1990. *Primate Visions: Gender, Race, and Nature in the World of Modern Science*. New York: Routledge.

Harding, Sandra. 1998. *Is Science Multicultural? Postcolonialisms, Feminisms, and Epistemologies*. Bloomington: Indiana University Press.

Harding, Sandra. 2004a. *The Feminist Standpoint Theory Reader: Intellectual and Political Controversies*. New York: Routledge.

Harding, Sandra. 2004b. "Introduction: Standpoint Theory as a Site of Political, Philosophic, and Scientific Debate." In *The Feminist Standpoint Theory Reader: Intellectual and Political Controversies*, ed. Sandra Harding, 1–16. New York: Routledge.

Harding, Sandra. 2006. *Science and Social Inequality: Feminist and Postcolonial Issues*. Urbana: University of Illinois Press.

Harding, Sandra. 2008. *Sciences from Below: Feminisms, Postcolonialities, and Modernities*. Durham, NC: Duke University Press.

Harley, J. B. 1989. "Deconstructing the Map." *Cartographica* 26:1–20.

Harley, J. B. 2001. *The New Nature of Maps: Essays in the History of Cartography*. Ed. Paul Laxton. Baltimore: Johns Hopkins University Press.

Harris, Leila M., and Helen D. Hazen. 2006. "Power of Maps: (Counter) Mapping for Conservation." *Acme* 4 (1): 99–130.

Harvey, David. 1997. *Justice, Nature, and the Geography of Difference*. Oxford: Blackwell.

Harvey, David. 2005. *Paris, Capital of Modernity*. 2nd ed. New York: Routledge.

Harvey, David. 2006. *Spaces of Global Capitalism: Towards a Theory of Uneven Geographical Development*. New York: Verso.

Harvey, Lee, and Morag MacDonald. 1993. *Doing Sociology: A Practical Introduction*. Houndmills, UK: Palgrave Macmillan.

Hashemite Kingdom of Jordan and State of Israel. 1998. *Treaty of Peace between the State of Israel and the Hashemite Kingdom of Jordan*. Vol. 2042, I–35325.

Hass, Amira. 2001. "Break-in at the Statistics Bureau." *Ha'aretz*, December 10.

Hass, Amira. 2011a. "Israel Allows Gaza Athletes to Cross into West Bank, but Bars Outstanding Academics." *Ha'aretz*, December 26.

Hass, Amira. 2011b. "West Bank Travel Restrictions Take Their Toll on International Aid Budgets." *Ha'aretz*, June 7.

Hass, Amira. 2012. "What Happens When a Palestinian Journalist Dares Criticize the Palestinian Authority?" *Ha'aretz*, April 2.

Hasson, Nir. 2011. "Israel Sanctions East Jerusalem Family for Straddling Palestinian Border." *Ha'aretz*, September 27.

Hasson, Nir. 2012a. "Google Street View Catches the Beauty and Hardships of Jerusalem's Old City." *Ha'aretz*, April 23.

Hasson, Nir. 2012b. "Rare Photograph Reveals Ancient Jerusalem Mosque Destroyed in 1967." *Ha'aretz*, June 15.

Hasson, Nir. 2013. "Caught on Tape: Recording Reveals East Jerusalem Park Is about Politics, Not Environment." *Ha'aretz*, September 30.

Hecht, Gabrielle. 2011. *Entangled Geographies: Empire and Technopolitics in the Global Cold War*. Cambridge, MA: MIT Press.

Henke, Christopher R., and Thomas F. Gieryn. 2008. "Sites of Scientific Practice: The Enduring Importance of Place." In *Handbook of Science and Technology Studies*, ed. Edward Hackett, Olga Amsterdamska, Michael E. Lynch, and Judy Wajcman, 353–375. Cambridge, MA: MIT Press.

Herod, Andrew, and Melissa W. Wright, eds. 2002. *Geographies of Power: Placing Scale*. Malden, MA: Blackwell.

Heruti-Sover, Tali. 2012. "Arab Town, Both Israeli and Palestinian, Divided by Shopping Israel News." *Ha'aretz*, February 1.

Hewitt, Rachel. 2010. *Map of a Nation: A Biography of the Ordnance Survey*. London: Granta.

Hirsch, Dafna. 2009. "We Are Here to Bring the West, Not Only to Ourselves? Zionist Occidentalism and the Discourse of Hygiene in Mandate Palestine." *International Journal of Middle East Studies* 41 (4): 577–594.

Hommels, Anique. 2005. *Unbuilding Cities: Obduracy in Urban Socio-Technical Change*. Cambridge, MA: MIT Press.

Hostetler, Laura. 2009. "Contending Cartographic Claims? The Qing Empire in Manchu, Chinese, and European Maps." In *The Imperial Map: Cartography and the Mastery of Empire*, ed. James R. Akerman, 93–132. Chicago: University of Chicago Press.

Houk, Marian. 2008. "Atarot and the Fate of the Jerusalem Airport." *Jerusalem Quarterly* 35:64–75.

Hull, Matthew S. 2003. "The File: Agency, Authority, and Autography in an Islamabad Bureaucracy." *Language and Communication* 23:287–314.

Human Rights Watch. 2008. "Off the Map: Land and Housing Rights Violations in Israel's Unrecognized Bedouin Villages (Excerpts)." *Journal of Palestine Studies* 37 (4): 176–178.

Ibn Khaldun, 'Abd al-Rahman. 1967. *The Muqaddimah: An Introduction to History*. Trans. Franz Rosenthal. Ed. and abridged N. J. Dawood. Princeton, NJ: Princeton University Press.

Ibrahim, Nassar. 2011. *Illusion of Development under Israeli Occupation: The Political Motivations of Donors to the Palestinians*. Bethlehem: Latin Patriarchate Printing Press.

"In Memory of Roberto Bachi." 1996. *Genus: Organo del Comitato Italiano per lo studio dei Problemi della Popolazione* 52 (1–2): 13–14.

Inbar, Moshe. 2002. "A Geomorphic and Environmental Evaluation of the Hula Drainage Project, Israel." *Australian Geographical Studies* 40 (2): 155–166.

Incite! Women of Color against Violence, ed. 2007. *The Revolution Will Not Be Funded: Beyond the Non-Profit Industrial Complex*. Cambridge, MA: South End Press.

InfoAmazonia. 2013. "About." http://infoamazonia.org/about (accessed September 19, 2013).

Ingham, Richard. 2013. "Cheers for Palestinian Film of Love and Betrayal." *Daily Star* (Beirut), May 20.

Ingold, Tim. 2007. *Lines: A Brief History*. London: Routledge.

Ingold, Tim. 2011. *Being Alive: Essays on Movement, Knowledge, and Description*. London: Routledge.

International Business Publications. 2011. *Israel Lobby in the United States Handbook: Organization, Operations, Performance*. Vol. 1. Washington, DC: International Business Publications.

International Peace and Cooperation Center. 2009. *Jerusalem, the Old City: The Urban Fabric and Geopolitical Implications*. Jerusalem: International Peace and Cooperation Center.

Isaac, Rami K. 2011. "Steadfastness and the Wall Conference in Bethlehem, Palestine." *Tourism Geographies* 13 (1): 152–157.

Israel Unveils Controversial Plan for West Bank Rail Network. 2013. *Al Akhbar English*, July 25.

Israeli Left Archive. n.d. "The Black Panthers in Israel." http://israeli-left-archive.org/cgi-bin/library?site=localhost&a=p&p=about&c=blackpan&l=en&w=utf-8 (accessed October 1, 2016).

Jasanoff, Sheila. 2004. *States of Knowledge: The Co-Production of Science and Social Order*. London: Routledge.

Jensen, Rolf H., Samih Abed, and Ulf Tellefsen. 1997. "Institution Building for Sustainable Physical Planning in Palestine." In *The Reconstruction of Palestine: Issues, Options, Policies, and Strategies*, ed. A. B. Zahlan, 76–85. London: Kegan Paul.

Jessop, Bob, and Stijn Oosterlynck. 2008. "Cultural Political Economy: On Making the Cultural Turn without Falling into Soft Economic Sociology." *Geoforum* 39 (3): 1155–1169.

Jiryis, Sabri, and Salah Qallab. 1985. "The Palestine Research Center." *Journal of Palestine Studies* 14 (4): 185–187.

Jones, Peris Sean. 2002. "The Etiquette of State-Building and Modernisation in Dependent States: Performing Stateness and the Normalisation of Separate Development in South Africa." *Geoforum* 33 (1): 25–40.

Jordan, June. 2002. *Some of Us Did Not Die: New and Selected Essays of June Jordan*. New York: Basic Books.

Jordanian Department of Lands and Surveys. 1966. *Annual Report: The Hashemite Kingdom of Jordan*. Uncataloged. Tel Aviv: Tel Aviv University Map Library.

Jubran, Hanna, Fawzi Nasser, and Farid Haj Yahya. 2008. *The Witnessing Roots: A Guidebook of Palestine*. Haifa: Arab Cultural Association.

Kagan, Oren. 1997. "The World of Scanning." Paper presented at EUROSTAT and the Israeli Central Bureau of Statistics workshop, New Technologies for the 2000 Census Round: Sharing the Israeli Experience from the 1995 Census, Kibbutz Ma'ale Hachamisha, Israel, March 16.

Kalpagam, U. 2000. "The Colonial State and Statistical Knowledge." *History of the Human Sciences* 13 (2): 37–55.

Kanaaneh, Rhoda Ann. 2002. *Birthing the Nation: Strategies of Palestinian Women in Israel*. Berkeley: University of California Press.

Kanaaneh, Rhoda Ann, and Isis Nusair, eds. 2010. *Displaced at Home: Ethnicity and Gender among Palestinians in Israel*. Albany: State University of New York Press.

Karsh, Efraim. 2000. *Fabricating Israeli History: The New Historians*. London: Frank Cass.

Karvonen, Andrew, and Bas Van Heur. 2014. "Urban Laboratories: Experiments in Reworking Cities." *International Journal of Urban and Regional Research* 38 (2): 379–392.

Katz, Cindi. 1994. "Playing the Field: Questions of Fieldwork in Geography." *Professional Geographer* 46 (1): 67–72.

Katz, Cindi. 2001. "On the Grounds of Globalization: A Topography for Feminist Political Engagement." *Signs* 26 (4): 1213–1234.

Katz, Cindi. 2006. "Messing with 'The Project.'" In *David Harvey: A Critical Reader*, ed. Noel Castree and Derek Gregory, 234–246. Oxford: Blackwell.

Katz, Cindi, and Andrew Kirby. 1991. "In the Nature of Things: The Environment and Everyday Life." *Transactions of the Institute of British Geographers* 16 (3): 259–271.

Katz, Kimberly. 2003. "Legitimizing Jordan as the Holy Land: Papal Pilgrimages—1964, 2000." *Comparative Studies of South Asia, Africa and the Middle East* 23 (1–2): 181–189.

Keane, Webb. 2007. *Christian Moderns: Freedom and Fetish in the Mission Encounter*. Berkeley: University of California Press.

Kedar, Benjamin Z. 1999. *The Changing Land between the Jordan and the Sea: Aerial Photographs from 1917 to the Present*. Jerusalem: Yad Ben-Zvi Press.

Keret, Etgar, and Assaf Hanuka. 2005. *Pizzeria Kamikaze*. Gainesville, FL: Alternative Comics.

Kestler-D'Amours, Jillian. 2013. "Bulldozers Flatten Bedouin Village 49 Times." *Al Jazeera English*, April 18.

Khalidi, Rashid. 2006. *The Iron Cage: The Story of the Palestinian Struggle for Statehood.* Boston: Beacon Press.

Khalidi, Rashid. 2010. *Palestinian Identity: The Construction of Modern National Consciousness.* New York: Columbia University Press.

Khalidi, Walid. 1988. "Plan Dalet: Master Plan for the Conquest of Palestine." *Journal of Palestine Studies* 18 (1): 4–33.

Khalidi, Walid. 1992. *All That Remains: The Palestinian Villages Occupied and Depopulated by Israel in 1948.* Washington, DC: Institute for Palestine Studies.

Khalidi, Walid. 1993. "Benny Morris and *Before Their Diaspora.*" *Journal of Palestine Studies* 22 (3): 106–119.

Khalidi, Walid. 2004. *Before Their Diaspora: A Photographic History of the Palestinians, 1876–1948.* Washington, DC: Institute for Palestine Studies.

Khalife, Marcel. 1980. *At the Border* ['Al-Hudud]. Beirut: CD Audio.

Khalili, Laleh. 2007. *Heroes and Martyrs of Palestine: The Politics of National Commemoration.* Cambridge: Cambridge University Press.

Khatib, Hisham. 2003. *Palestine and Egypt under the Ottomans: Paintings, Books, Photographs, Maps, and Manuscripts.* London: Tauris Parke.

Khoury, Elias. 2005. *Gate of the Sun* [Bab al-Shams]. Trans. Humphrey Davies. Brooklyn: Archipelago Books.

Kitchin, Rob, and Martin Dodge. 2007. "Rethinking Maps." *Progress in Human Geography* 31 (3): 331–344.

Kivelson, Valerie A. 2009. "'Exalted and Glorified to the Ends of the Earth': Imperial Maps and Christian Spaces in Seventeenth- and Early Eighteenth-Century Russian Siberia." In *The Imperial Map: Cartography and the Mastery of Empire,* ed. James R. Akerman, 47–92. Chicago: University of Chicago Press.

Klibanoff, Lea, dir. 2009. *Ha-Mashiach Tamid Yavo* [The Messiah Will Always Come]. Documentary.

Knorr Cetina, Karin. 1981. *The Manufacture of Knowledge: An Essay on the Constructivist and Contextual Nature of Science.* Oxford: Pergamon Press.

Knorr Cetina, Karin. 1999. *Epistemic Cultures: How the Sciences Make Knowledge.* Cambridge, MA: Harvard University Press.

Koeppel, Gerard. 2015. *City on a Grid: How New York Became New York.* Boston: Da Capo Press.

Kohler, Robert E. 2002. *Landscapes and Labscapes: Exploring the Lab–Field Border in Biology.* Chicago: University of Chicago Press.

Kotef, Hagar. 2015. *Movement and the Ordering of Freedom: On Liberal Governances of Mobility*. Durham, NC: Duke University Press.

Kouw, Matthijs, Charles van den Heuvel, and Andrea Scharnhorst. 2013. "Exploring Uncertainty in Knowledge Representations: Classifications, Simulations, and Models of the World." In *Virtual Knowledge: Experimenting in the Humanities and the Social Sciences*, ed. Paul Wouters, Anne Beaulieu, Andrea Scharnhorst, and Sally Wyatt, 89–126. Cambridge, MA: MIT Press.

Kreimer, Sarah. 2007. *A Policy Framework for the Interim Period*. Jerusalem: Peace and Democracy Forum.

Kuhn, Thomas S. 1996. *The Structure of Scientific Revolutions*. Chicago: University of Chicago Press.

Kurgan, Laura. 2013. *Close Up at a Distance: Mapping, Technology, and Politics*. Brooklyn: Zone Books.

Kwan, Mei-Po. 2004. "Beyond Difference: From Canonical Geography to Hybrid Geographies." *Annals of the Association of American Geographers* 94 (4): 756–763.

Larkin, Craig. 2014. "Jerusalem's Separation Wall and Global Message Board: Graffiti, Murals, and the Art of Sumud." *Arab Studies Journal* 22 (1): 134–169.

Lasman, Benjamin. 1997. "Overview of the 1995 Census of Population and Housing in Israel." Paper presented at EUROSTAT and the Israeli Central Bureau of Statistics workshop, New Technologies for the 2000 Census Round: Sharing the Israeli Experience from the 1995 Census, Kibbutz Ma'ale Hachamisha, Israel, March 16.

Latour, Bruno. 1986. "Visualization and Cognition." *Knowledge in Society* 6:1–40.

Latour, Bruno. 1987. *Science in Action: How to Follow Scientists and Engineers through Society*. Cambridge, MA: Harvard University Press.

Latour, Bruno. 1988. *The Pasteurization of France*. Cambridge, MA: Harvard University Press.

Latour, Bruno. 1993. *We Have Never Been Modern*. Cambridge, MA: Harvard University Press.

Latour, Bruno. 1999. "On Recalling ANT." In *Actor Network Theory and After*, ed. John Law and John Hassard, 15–25. Oxford: Blackwell.

Latour, Bruno. 2005. *Reassembling the Social: An Introduction to Actor-Network-Theory*. Oxford: Oxford University Press.

Latour, Bruno, and Emilie Hermant. 1998. *Paris ville invisible* [Paris: Invisible City]. Paris: La Découverte et Institut Synthélabo pour le Progrès de la Connaissance.

Latour, Bruno, and Steve Woolgar. 1986. *Laboratory Life: The Construction of Scientific Facts*. Princeton, NJ: Princeton University Press.

Law, John, and Annemarie Mol. 2001. "Situating Technoscience: An Inquiry into Spatialities." *Environment and Planning D: Society and Space* 19 (5): 609–621.

Law, John, Evelyn Ruppert, and Mike Savage. 2010. "The Double Social Life of Methods." Centre for Research on Socio-Cultural Change, Working Paper Series. London: Goldsmiths Research Online.

Lee, Benjamin, and Edward LiPuma. 2002. "Cultures of Circulation: The Imaginations of Modernity." *Public Culture* 14 (1): 191–213.

Lefebvre, Henri. 2002. *The Production of Space*. Oxford: Blackwell.

Leibler, Anat. 2004. "Statisticians' Ambition: Governmentality, Modernity, and National Legibility." *Israel Studies* 9 (2): 121–149.

Leibler, Anat. 2007. "Establishing Scientific Authority: Citizenship and the First Census of Israel." In *Tel Aviver Jahrbuch für deutsche Geschichte XXXV*, ed. Jos. Brunner, 221–236. Göttingen: Wallstein Verlag.

Leibler, Anat, and Daniel Breslau. 2005. "The Uncounted: Citizenship and Exclusion in the Israeli Census of 1948." *Ethnic and Racial Studies* 28 (5): 880–902.

Leuenberger, Christine. 2012. "Map-Making for Palestinian State-Making." *Arab World Geographer* 16 (1): 54–74.

Leuenberger, Christine, and Izhak Schnell. 2010. "The Politics of Maps: Constructing National Territories in Israel." *Social Studies of Science* 40 (3): 803–842.

Levin, Noam. 2006. "The Palestine Exploration Fund Map (1871–1877) of the Holy Land as a Tool for Analysing Landscape Changes: The Coastal Dunes of Israel as a Case Study." *Cartographic Journal* 43 (1): 45–67.

Levin, Noam, Eldad Elron, and Avital Gasith. 2009. "Decline of Wetland Ecosystems in the Coastal Plain of Israel during the 20th Century: Implications for Wetland Conservation and Management." *Landscape and Urban Planning* 92 (3–4): 220–232.

Levinson, Meira. 2007. "Making the Desert Bloom: Israel's Environmental Past and Zionist Future." *PresenTense* 5:n.p.

Livingstone, David N. 1992. *The Geographical Tradition*. Oxford: Blackwell.

Livingstone, David N. 2003. *Putting Science in Its Place: Geographies of Scientific Knowledge*. Chicago: University of Chicago Press.

Longley, Paul A., and Michael Batty. 2003. *Advanced Spatial Analysis: The CASA Book of GIS*. Redlands, CA: ESRI Press.

Loolwa, Khazoom. 2013. "Jews of the Middle East." *Jewish Virtual Library*, May 11.

Lorde, Audre. 2007. *Sister Outsider: Essays and Speeches*. Berkeley, CA: Crossing Press.

Louder, Dean R., Michel Bisson, and Pierre La Rochelle. 1974. "Analyse Cen-trographique de la Population du Québec de 1951 à 1971" [Centrographic Analysis of the Population of Quebec from 1951 to 1971]. *Cahiers de Geographie de Quebec* 18 (45): 421–444.

Low, Setha M., and Denise Lawrence-Zúñiga. 2003. *The Anthropology of Space and Place: Locating Culture*. Hoboken, NJ: Wiley-Blackwell.

Lowdermilk, Walter C. 1960. "The Reclamation of a Man-Made Desert." *Scientific American*, April.

Lynch, Michael E. 1993. *Scientific Practice and Ordinary Action: Ethnomethodology and Social Studies of Science*. Cambridge: Cambridge University Press.

Mackenzie, Donald A. 1990. *Inventing Accuracy: A Historical Sociology of Nuclear Missile Guidance*. Cambridge, MA: MIT Press.

MacKenzie, Donald A., and Judy Wajcman, eds. 1999. *The Social Shaping of Technology: How the Refrigerator Got Its Hum*. Philadelphia: Open University Press.

Macleod, Roy M., and Deepak Kumar. 1995. *Technology and the Raj: Western Technology and Technical Transfers to India, 1700–1947*. New Delhi: Sage.

Mahmood, Saba. 2005. *Politics of Piety: The Islamic Revival and the Feminist Subject*. Princeton,NJ: Princeton University Press.

Ma'oz, Moshe, and Sari Nusseibeh, eds. 2000. *Jerusalem: Points of Friction and Beyond*. The Hague: Kluwer Law International.

"Map of Palestine." 2015.http://www.mapofpalestine.com (accessed September 21, 2015).

Margalit, Meir. 2010. *Seizing Control of Space in East Jerusalem*. Jerusalem: Sifrei Aliat Gag.

Marie, Mohammad. 2015. "Resilience of Nurses Who Work in Community Mental Health Workplaces in West Bank-Palestine." PhD diss., Cardiff University.

Markell, Patchen. 2003. *Bound by Recognition*. Princeton, NJ: Princeton University Press.

Marston, Sallie A. 2000. "The Social Construction of Scale." *Progress in Human Geography* 24 (2): 219–242.

Marston, Sallie A., John P. Jones, and Keith Woodward. 2005. "Human Geography without Scale." *Transactions of the Institute of British Geographers* 30 (4): 416–432.

Martin, Enrique Martino. 2011. "Largely a Trojan Horse: Provincializing the Scale Debate in the Political Economy of Globalisation." Paper presented at the Colonial Legacies, Postcolonial Contestations: Decolonizing the Social Sciences and the

Humanities conference, Frankfurt Research Center for Postcolonial Studies, June 16–18 June.

Masalha, Nur, ed. 2005. *Catastrophe Remembered: Palestine, Israel, and the Internal Refugees. Essays in Memory of Edward W. Said.* London: Zed Books.

Massad, Joseph Andoni. 2001. *Colonial Effects: The Making of National Identity in Jordan.* New York: Columbia University Press.

Massad, Joseph Andoni. 2003. "The Ends of Zionism: Racism and the Palestinian Struggle." *Interventions* 5 (3): 440–451.

Massad, Joseph Andoni. 2006. *The Persistence of the Palestinian Question: Essays on Zionism and the Palestinians.* New York: Routledge.

Massey, Doreen. 1994. *Space, Place, and Gender.* Minneapolis: University of Minnesota Press.

Massey, Doreen. 2005. *For Space.* London: Sage.

Matar, Haggai. 2012. "Watch: Three Palestinian NGO Offices Raided by IDF Overnight." *+972 Magazine,* December 11.

Matar, Haggai. 2016. "The End of History at Israel's State Archives?" *+972 Magazine,* April 12.

Meari, Lena Mhammad. 2011. "Sumud: A Philosophy of Confronting Interrogation." PhD diss., University of California at Davis.

Meari, Lena Mhammad. 2014. "Sumud: A Palestinian Philosophy of Confrontation in Colonial Prisons." *South Atlantic Quarterly* 113 (3): 547–578.

Meir-Glitzenstein, Esther. 2011. "Operation Magic Carpet: Constructing the Myth of the Magical Immigration of Yemenite Jews to Israel." *Israel Studies* 16 (3): 149–173.

Mermelstein, Hannah. 2011. "Overdue Books: Returning Palestine's 'Abandoned Property' of 1948." *Jerusalem Quarterly* 47:46–64.

Merrifield, Andy. 2000. "Henri Lefebvre: A Socialist in Space." In *Thinking Space*, ed. Mike Crang and Nigel Thrift, 167–182. London: Routledge.

MIFTAH. 2010. "Palestinian Avatars Protest Israeli Measures." MIFTAH, February 13.

Military Court Watch. 2013. "Two Boys, Two Laws: The Discriminatory Application of Law in the West Bank." http://www.militarycourtwatch.org/files/server/UN%20 SUBMISSION%20-%20DISCRIMINATION.pdf (accessed August 17, 2016).

Miller, Anna Lekas. 2013. "'No Falasteen for You!' Shin Bet Banned Me from Israel for 10 Years." *+972 Magazine,* August 31.

Ministry of Defense. 1973. "The Administered Territories 1972/73: Data on Civilian Activities in Judea and Samaria, the Gaza Strip, and Northern Sinai." Israeli Ministry of Defense, Coordinator of Government Operations in the Administered Territories.

Misselwitz, Philipp, and Tim Rieniets, eds. 2006. *City of Collision: Jerusalem and the Principles of Conflict Urbanism.* Basel: Birkhauser.

Mitchell, Andrew P. 1948. "Supplement to 'Report for 1940–1946' for the Period up to [March 31, 1948]." In *Report for the Years 1940–1946.* Jerusalem: British Mandate SOP.

Mitchell, Timothy. 1991. *Colonising Egypt.* Berkeley: University of California Press.

Mitchell, Timothy. 2002. *Rule of Experts: Egypt, Techno-Politics, Modernity.* Berkeley: University of California Press.

Mitchell, Timothy. 2008. "Rethinking Economy." *Geoforum* 39 (3): 1116–1121.

Mitchell, Timothy. 2014. "Introduction: Life of Infrastructure." *Comparative Studies of South Asia, Africa, and the Middle East* 34 (3): 437–439.

Mitchell, W.J.T., ed. 2002. *Landscape and Power.* Chicago: University of Chicago Press.

Modan, Rutu. 2008. *Exit Wounds.* Montreal: Drawn and Quarterly.

Mogel, Lize and Alexis Bhagat, eds. 2007. *An Atlas of Radical Cartography.* Los Angeles: Journal of Aesthetics and Protest Press.

Mohanty, Chandra Talpade, Ann Russo, and Lourdes Torres, eds. 1991. *Third World Women and the Politics of Feminism.* Bloomington: Indiana University Press.

Mol, Annemarie. 2002. *The Body Multiple: Ontology in Medical Practice.* Durham, NC: Duke University Press.

Monmonier, Mark S. 1991. *How to Lie with Maps.* Chicago: University of Chicago Press.

Mood, Fulmer. 1946. "The Rise of Official Statistical Cartography in Austria, Prussia, and the United States, 1855–1872." *Agricultural History* 20 (4): 209–225.

Moore, Kate. 2007. "Towards a Post-Colonial GIS." In *GIS Research UK (GISRUK) 2007 Proceedings,* ed. Adam C. Winstanley. Maynooth: National University of Ireland.

MOPIC (Ministry of Planning and International Cooperation). 1996. *Landscape Assessment of the West Bank Governorates: Emergency Natural Resources Protection Plan.* Ramallah: Palestinian National Authority MOPIC, Directorate for Urban and Rural Planning.

MOPIC. 1998. *The Regional Plan for the West Bank Governorates.* Ramallah: MOPIC.

Morris, Benny. 1992. "Palestine to 1948." *Journal of Palestine Studies* 22 (1): 109–111.

Morris, Benny. 2004. *The Birth of the Palestinian Refugee Problem Revisited*. Cambridge: Cambridge University Press.

Mrázek, Rudolf. 2002. *Engineers of Happy Land: Technology and Nationalism in a Colony*. Princeton, NJ: Princeton University Press.

Munayyer, Yousef. 2012. "A State That Fears the Womb." *Daily Beast*, April 5.

Nagra, Ruhan. 2013. "Academia Undermined: Israeli Restrictions on Foreign National Academics in Palestinian Higher Education Institutions." Ramallah: Right to Enter.

Nakba Archive. n.d. http://nakba-archive.org (accessed July 15, 2015).

Nassar, Issam. 2003. "Remapping Palestine and the Palestinians: Decolonizing and Research." *Comparative Studies of South Asia, Africa, and the Middle East* 23 (1–2): 149–151.

Nasser, Irene. 2013. "In Bab Al-Shams, Palestinians Create New Facts on the Ground." *+972 Magazine*, January 25.

Natour, Salman, and Yusuf Abu Ta'ah. 2000. *Filistin 'ala al-Tariq min al-Nasira ila Bayt Laham* [Palestine on the Road: From Nazareth to Bethlehem]. Ramallah: Tamer Institute for Community Education.

Nayar, Pramod K. 2012. *Colonial Voices: The Discourses of Empire*. Chichester, UK: Wiley-Blackwell.

Newman, David. 2001. "From National to Post-National Territorial Identities in Israel-Palestine." *GeoJournal* 53 (3): 235–246.

"NGO Monitor." 2015. *Wikipedia*. September 30. https://en.wikipedia.org/wiki/NGO_Monitor (accessed December 16, 2015).

Nitzan, Jonathan, and Shimshon Bichler. 2002. *The Global Political Economy of Israel*. London: Pluto Press.

November, Valérie, Eduardo Camacho-Hübner, and Bruno Latour. 2010. "Entering a Risky Territory: Space in the Age of Digital Navigation." *Environment and Planning D: Society and Space* 28 (4): 581–599.

Noy, Chaim. 2008. "Writing Ideology: Hybrid Symbols in a Commemorative Visitor Book in Israel." *Journal of Linguistic Anthropology* 18 (1): 62–81.

Nusseibeh, Jamal. 2012. "Israeli Army Invades and Closes University Community Media Center." Open letter. Al-Quds University, April 2.

Nayel, Moe Ali. 2013. "Palestinian Refugees Are Not at Your Service." *Electronic Intifada*, May 17.

Ofran, Hagit, Jad Isaac, Hillel Ben-Sasson, Omar Yousef, and Adnan Abdelrazek. 2011. "Talking about Jerusalem." Paper presented at the *Palestine–Israel Journal*'s Jerusalem: In the Eye of the Storm conference, Jerusalem, April 13.

Ophir, Adi, Michal Givoni and Sari Hanafi, eds. 2009. *The Power of Inclusive Exclusion: Anatomy of Israeli Rule in the Occupied Palestinian Territories*. New York: Zone Books.

Orient House. 2001a. "Israel Continues Its Efforts to Strangle the Orient House." Orient House, July 20.

Orient House. 2001b. "The Looted Archives of the Orient House." *Jerusalem Quarterly* 13:3–5.

Orient House. 2001c. "Orient House Begins Its Commemoration Ceremony despite Israeli Obstacles." Orient House, July 17.

Orient House. 2001d. "Press Release from the Orient House." Orient House, August 30.

Orient House. 2001e. "Urgent Appeal from the Orient House." Orient House, August 14.

Orient House. 2005. Statement from the Orient House to the International Community. Orient House, July 19.

Oster, Marcy. 2011. "Critics of Knesset NGO Bills Say Israel's Democracy Being Undermined." Jewish Telegraphic Agency, November 14.

O'Sullivan, David. 2006. "Geographical Information Science: Critical GIS." *Progress in Human Geography* 30 (6): 783–791.

Paasi, Anssi. 2004. "Place and Region: Looking through the Prism of Scale." *Progress in Human Geography* 28 (4): 536–546.

PACBI (Palestinian Campaign for the Academic and Cultural Boycott of Israel). 2009. "Guidelines for the International Academic Boycott of Israel." October.

PACBI. 2011. "Israel's Exceptionalism: Normalizing the Abnormal." October.

Paglen, Trevor. 2009. *Blank Spots on the Map: The Dark Geography of the Pentagon's Secret World*. New York: Dutton.

"Palestinian Economy Minister Charged with Embezzlement, Insider Trading." 2011. *Ha'aretz*, November 29.

PalMap. 2011. *Jerusalem Street Atlas*. Bethlehem: Palestine Mapping Center.

Pappé, Ilan. 1992. *The Making of the Arab–Israeli Conflict, 1947–51*. London: I. B. Tauris.

Paraszczuk, Joanna. 2011. "NGOs: Bill Is a 'Shameful Moment' for Israel." *Jerusalem Post*, November 13.

Parker, Christopher. 1999. *Resignation or Revolt: Socio-Political Development and the Challenges of Peace in Palestine*. London: I. B. Tauris.

Parsons, Nigel. 2011. "The Palestinian Authority Security Apparatus: Biopolitics, Surveillance, and Resistance in the Occupied Palestinian Territories." In *Surveillance and Control in Israel/Palestine: Population, Territory, and Power*, ed. Elia Zureik, David Lyon, and Yasmeen Abu-Laban, 355–370. London: Routledge.

Pavlovskaya, Marianna. 2006. "Theorizing with GIS: A Tool for Critical Geographies?" *Environment and Planning A* 38 (11): 2003–2020.

Pavlovskaya, Marianna, and Jess Bier. 2012. "Mapping Census Data for Difference: Towards the Heterogeneous Geographies of Arab American Communities of the New York Metropolitan Area." *Geoforum* 43 (3): 483–496.

PCBS (Palestinian Central Bureau of Statistics). 2008. "Press Conference on the Preliminary Findings (Population, Housing, and Establishment Census 2007)." Ramallah: PCBS.

Peace Now. 2007. "Guilty! Construction of Settlements upon Private Land: Official Data." Jerusalem: Peace Now Settlement Watch.

Peace Now. 2011. "West Bank and Jerusalem Map, the Settlements: The Biggest Threat to a Two-State Solution." Jerusalem: Peace Now Settlement Watch.

Peace Now. 2012. "E1 Area: Planning Status." Jerusalem: Peace Now Settlement Watch.

Peet, Richard. 1985. "The Social Origins of Environmental Determinism." *Annals of the Association of American Geographers* 75 (3): 309–333.

Peled, Ammatzia. 1996. "Generating the Israeli National GIS." In *The Mosaic of Israeli Geography*, ed. Y. Gradus and Gabriel Lipshitz, 485–496. Beer Sheva: Ben-Gurion University of the Negev Press.

Perlmann, Joel. 2012. *The 1967 Census of the West Bank and Gaza Strip: A Digitized Version*. Annandale-on-Hudson, NY: Levy Economics Institute of Bard College. http://www.levyinstitute.org/palestinian-census (accessed October 21, 2015).

Peters, Peter, Sanneke Kloppenburg, and Sally Wyatt. 2010. "Co-ordinating Passages: Understanding the Resources Needed for Everyday Mobility." *Mobilities* 5 (3): 349–368.

Pickles, John, ed. 1995. *Ground Truth: The Social Implications of Geographic Information Systems*. New York: Guilford Press.

Pickles, John, ed. 2004. *A History of Spaces: Cartographic Reason, Mapping, and the Geo-Coded World*. New York: Routledge.

Pinch, Trevor, and Wiebe Bijker. 2012. "The Social Construction of Facts and Artefacts: Or, How the Sociology of Science and the Sociology of Technology Might Benefit Each Other." In *The Social Construction of Technological Systems*, ed. Wiebe Bijker, Thomas P. Hughes and Trevor Pinch, 11–44. Cambridge, MA: MIT Press.

Piper, Karen. 2002. *Cartographic Fictions: Maps, Race, and Identity*. New Brunswick, NJ: Rutgers University Press.

PLO and State of Israel. 1995. *Annex I: Protocol concerning Redeployment and Security Arrangements* [Oslo II Agreement between the Palestinian Liberation Organization (PLO) and Israel].

Porter, Theodore M., and Dorothy Ross, eds. 2003. *The Cambridge History of Science*. Vol. 7. Cambridge: Cambridge University Press.

Povinelli, Elizabeth. 2006. *The Empire of Love: Toward a Theory of Intimacy, Genealogy, and Carnality*. Durham, NC: Duke University Press.

Pratt, Mary Louise. 2007. *Imperial Eyes: Travel Writing and Transculturation*. London: Routledge.

Pullan, Wendy. 2013. "Conflict's Tools: Borders, Boundaries, and Mobility in Jerusalem's Spatial Structures." *Mobilities* 8 (1): 125–147.

Pullan, Wendy, and Britt Baillie, eds. 2013. *Locating Urban Conflicts: Ethnicity, Nationalism, and the Everyday*. Houndmills, UK: Palgrave Macmillan.

Qarmout, Tamer, and Daniel Beland. 2012. "The Politics of International Aid to the Gaza Strip." *Journal of Palestine Studies* 41 (4): 1–16.

Qato, Mezna. 2016. "Returns of the Archive." *Nakba Files*, June 1.

Quiquivix, Linda. 2013. "When the Carob Tree Was the Border: On Autonomy and Palestinian Practices of Figuring It Out." *Capitalism, Nature, Socialism* 24 (3): 170–189.

Radcliffe, Sarah A. 2009. "National Maps, Digitalisation, and Neoliberal Cartographies: Transforming Nation-State Practices and Symbols in Postcolonial Ecuador." *Transactions of the Institute of British Geographers* 34 (4): 426–444.

Radder, Hans. 1996. *In and About the World: Philosophical Studies of Science and Technology*. Albany: State University of New York Press.

Radder, Hans. 2006. *The World Observed, the World Conceived*. Pittsburgh: University of Pittsburgh Press.

Raj, Kapil. 2007. *Relocating Modern Science: Circulation and the Construction of Knowledge in South Asia and Europe, 1650–1900*. Houndmills, UK: Palgrave Macmillan.

Rappert, Brian. 2009. *Experimental Secrets: International Security, Codes, and the Future of Research*. Lanham, MD: University Press of America.

Rashi Foundation. 2012. "Funding the Future: Advancing STEM in Israeli Education." Conference summary. New York: Rashi Foundation and Jewish Funders Network.

Razin, Assaf. 1993. *The Economy of Modern Israel: Malaise and Promise*. Chicago: University of Chicago Press.

Redfield, Peter. 2012. "The Unbearable Lightness of Ex-Pats: Double Binds of Humanitarian Mobility." *Cultural Anthropology* 27 (2): 358–382.

Ricca, Simone. 2007. *Reinventing Jerusalem: Israel's Reconstruction of the Jewish Quarter after 1967*. London: I. B. Tauris.

Robbins, Annie, and Phil Weiss. 2013. "'Jewish Settlements on West Bank' Are Now Comedy Central Fare." Mondoweiss, August 15. http://mondoweiss.net/2013/08/jewish-settlements-on-west-bank-are-now-comedy-central-fare (accessed November 2, 2016).

Rodgers, Dennis, and Bruce O'Neill. 2012. "Infrastructural Violence: Introduction to the Special Issue." *Ethnography* 13 (4): 401–412.

Rogers, Richard, and Anat Ben-David. 2008. "The Palestinian–Israeli Peace Process and Transnational Issue Networks: The Complicated Place of the Israeli NGO." *New Media and Society* 10 (3): 497–528.

Romani, Vincent. 2008. "Sciences sociales et lutte nationale dans les Territoires Occupés Palestiniens: La coercition comme contrainte et comme ressource scientifique" [Social Science and National Struggle in the Occupied Palestinian Territories: Coercion as a Constraint and as a Scientific Resources]. *Revue d'Anthropologie des Connaissances* 2 (3): 487–504.

Rose, Gillian. 1992. "Geography as the Science of Observation: The Landscape, the Gaze, and Masculinity." In *Nature and Science: Essays in the History of Geographical Knowledge*, ed. Felix Driver and Gillian Rose, 8–18. Lancaster: Historical Geography Research Group of the Institute of British Geographers.

Rose, Gillian. 1997. "Situating Knowledges: Positionality, Reflexivities, and Other Tactics." *Progress in Human Geography* 21 (3): 305–320.

Rose, Mitch, and John Wylie. 2006. "Animating Landscape." *Environment and Planning D: Society and Space* 24 (4): 475–479.

Rosen, Rhoda. 2008. *Imaginary Coordinates*. Chicago: Spertus Press.

Rosenberg, Oz. 2012. Peace Now Activist's Home Vandalized for the Third Time in a Year. *Ha'aretz*, July 16.

Rosenblum, Keshet. 2012. "Jerusalem Official Demands 'Zionist Architect' for National Library." *Ha'aretz*, November 6.

Rotbard, Sharon. 2015. *White City, Black City: Architecture and War in Tel Aviv and Jaffa*. Cambridge, MA: MIT Press.

Sacco, Joe. 2001. *Palestine*. Seattle: Fantagraphics Books.

Sacco, Joe. 2009. *Footnotes in Gaza*. New York: Metropolitan Books.

Safian, Alex. 2007. "Peace Now's Blunder: Erred on Ma'ale Adumim Land by 15,900 Percent." Committee for Accuracy in Middle East Reporting in America.

Safieddine, Hicham. 2004. "Sami Hadawi, 100: Canadian Palestinian Scholar." *Toronto Star*, November 25.

Safier, Neil. 2009. "The Confines of the Colony: Boundaries, Ethnographic Landscapes, and Imperial Cartography in Iberoamerica." In *The Imperial Map: Cartography and the Mastery of Empire*, ed. James R. Akerman, 133–184. Chicago: University of Chicago Press.

Said, Edward W. 1983. "Traveling Theory." In *The World, the Text, and the Critic*, 226–247. Cambridge, MA: Harvard University Press.

Said, Edward W. 1994. *Culture and Imperialism*. London: Vintage Books.

Said, Edward W. 1995. *Peace and Its Discontents: Essays on Palestine in the Middle East Peace Process*. New York: Vintage Books.

Said, Edward W. 2001. *The End of the Peace Process: Oslo and After*. New York: Knopf Doubleday.

Salamanca, Omar Jabary. 2011. "Unplug and Play: Manufacturing Collapse in Gaza." *Human Geographies* 4 (1): 22–37.

Salamanca, Omar Jabary. 2014. "Fabric of Life: The Infrastructure of Settler Colonialism and Uneven Development in Palestine." PhD diss., Ghent University.

Salamanca, Omar Jabary, Mezna Qato, Kareem Rabie, and Sobhi Samour. 2012. "Past Is Present: Settler Colonialism in Palestine." *Settler Colonial Studies* 2 (1): 1–8.

Saleh, Ahmad. 2008. "Reshaping Palestinian Urban Structure towards Sustainable Urban Development." Paper presented at the First International Conference on Urban Planning in Palestine: Current Challengers and Future Prospects, Nablus, October 21.

Saloul, Ihab. 2012. *Catastrophe and Exile in the Modern Palestinian Imagination: Telling Memories*. Houndmills, UK: Palgrave Macmillan.

Sauer, Carl. 1918. "Geography and the Gerrymander." *American Political Science Review* 12 (3): 403–426.

Sauer, Carl. 1925. *The Morphology of Landscape*. Berkeley: University of California Press.

Sayegh, Faris. 2000. *Geographic Data Survey, Needs Analysis, and Implementation Plan*. Annex 1. Negotiations Support Unit, Ramallah, West Bank. Physical Planning Box 1. ARIJ Library.

Sayigh, Rosemary. 1977. "The Palestinian Identity among Camp Residents." *Journal of Palestine Studies* 6 (3): 3–22.

Schaffer, Simon. 1994. *From Physics to Anthropology—and Back Again*. Cambridge, UK: Prickly Pear Press.

Scham, Sandra Arnold. 2003. "'From the River unto the Land of the Philistines': The 'Memory' of Iron Age Landscapes in Modern Visions of Palestine." In *Deterritorialisations ... Revisioning: Landscapes and Politics*, ed. Mark Dorrian and Gillian Rose, 73–79. London: Black Dog.

Schinkel, Willem. 2010. *Aspects of Violence: A Critical Theory*. Houndmills, UK: Palgrave Macmillan.

Schiocchet, Leonardo. 2011. *Palestinian Sumud: Steadfastness, Ritual, and Time among Palestinian Refugees*. Scholarly Paper 2130405. Rochester, NY: Social Science Research Network.

Schivelbusch, Wolfgang. 1987. "The Policing of Street Lighting." *Yale French Studies* 73:61–74.

Schmelz, Usiel O., and Nathan Gad, eds. 1986. *Studies in the Population of Israel in Honor of Roberto Bachi. Scripta Hierosolymitana XXX*. Jerusalem: Magnes Press.

Scholten, Henk J., Rob van de Velde, and Niels van Manen, eds. 2009. *Geospatial Technology and the Role of Location in Science*. Dordrecht: Springer.

Schuurman, Nadine. 1999. "Critical GIS: Theorizing an Emerging Discipline." *Cartographica* 36 (4): 1–101.

Schuurman, Nadine. 2000. "Trouble in the Heartland: GIS and Its Critics in the 1990s." *Progress in Human Geography* 24 (4): 569–590.

Scott, Heidi V. 2006. "Rethinking Landscape and Colonialism in the Context of Early Spanish Peru." *Environment and Planning D: Society and Space* 24 (4): 481–496.

Scott, James C. 1998. *Seeing Like a State: How Certain Schemes to Improve the Human Condition Have Failed*. New Haven, CT: Yale University Press.

Segal, Rafi, and Eyal Weizman, eds. 2003a. *A Civilian Occupation: The Politics of Israeli Architecture*. London: Verso.

Segal, Rafi, and Eyal Weizman. 2003b. "Occupation in Space and Time." *Index on Censorship* 32 (3): 186–192.

Segev, Tom. 2007. *1967: Israel, the War, and the Year That Transformed the Middle East*. Trans. Jessica Cohen. New York: Henry Holt.

Seikaly, May. 1995. *Haifa: Transformation of a Palestinian Arab Society, 1918–1939*. London: I. B. Tauris.

Sekhsaria, Pankaj. 2011. "Jugaad as a Conceptual and Materials Commons." *Common Voices* 8:21–23.

Sekhsaria, Pankaj. 2013. "The Making of an Indigenous Scanning Tunneling Microscope." *Current Science* 104 (9): 1152–1158.

Selmeczi, Anna. 2009. "'… We Are Being Left to Burn Because We Do Not Count': Biopolitics, Abandonment, and Resistance." *Global Society* 23 (4): 519–538.

Senor, Dan, and Saul Singer. 2009. *Start-up Nation: The Story of Israel's Economic Miracle*. New York: Twelve.

Shafak, Elif. 2005. "Linguistic Cleansing." *New Perspectives Quarterly* 22 (3): 19–25.

Shahin, Mariam. 2005. *Palestine: A Guide*. Northampton, MA: Interlink Books.

Shalev, Nir. 2012. *Under the Guise of Legality: Israel's Declarations of State Land in the West Bank*. Trans. Zvi Shulman. Jerusalem: B'Tselem.

Shamir, Ronen. 1996. "Suspended in Space: Bedouins under the Law of Israel." *Law and Society Review* 30 (2): 231–257.

Shamsavari, Ali. 2007. "The Technology Transfer Paradigm: A Critique." Discussion paper. Kingston-upon-Thames, UK: Kingston University.

Shapin, Steven and Simon Schaffer. 1985. *Leviathan and the Air-Pump: Hobbes, Boyle, and the Experimental Life*. Princeton, NJ: Princeton University Press.

Sharakas, Othman, Ahmad Nubani, Ahmad Abu Hammad, and Abdullah Abdullah. 2007. *Land Degradation Risk Assessment in the Palestinian Central Mountains Utilizing Remote Sensing and GIS Technique*. Lower Jordan River Basin Programme Publications 14. Ramallah: Birzeit University.

Sharp, Joanne P. 2009. *Geographies of Postcolonialism: Spaces of Power and Representation*. Los Angeles: Sage.

Shea, K. Stuart, and Robert B. McMaster. 1989. "Cartographic Generalization in a Digital Environment: When and How to Generalize." *Autocarto* 9:56–67.

Shehadeh, Raja. 1982. *The Third Way: A Journal of Life in the West Bank*. London: Quartet Books.

Shehadeh, Raja. 2007. *Palestinian Walks: Forays into a Vanishing Landscape*. New York: Scribner.

Shehadeh, Raja. 2013. "Stones Unturned." *New York Times*, June 20.

Sheizaf, Noam. 2012. "Google Street View to Feature West Bank Settlements." *+972 Magazine,* September 27.

Sheizaf, Noam. 2013. "Stephen Hawking's Message to Israeli Elites: The Occupation Has a Price." *+972 Magazine,* May 8.

Sheller, Mimi, and John Urry. 2006. "The New Mobilities Paradigm." *Environment and Planning A* 38 (2): 207–226.

Shenhav, Yehouda, and Yael Berda. 2009. "The Colonial Foundations of the State of Exception: Juxtaposing the Israeli Occupation of the Palestinian Territories with Colonial Bureaucratic History." In *The Power of Inclusive Exclusion: Anatomy of Israeli Rule in the Occupied Palestinian Territories,* ed. Adi Ophir, Michal Givoni and Sari Hanafi, 337–374. New York: Zone Books.

Sherstyuk, Katerina. 1998. "How to Gerrymander: A Formal Analysis." *Public Choice* 95 (1–2): 27–49.

Shohat, Ella. 1992. "Antimonies of Exile: Said at the Frontiers of National Narrations." In *Edward Said: A Critical Reader,* ed. Michael Sprinker, 121–143. Oxford: Blackwell.

Shohat, Ella. 2002. "Area Studies, Gender Studies, and the Cartographies of Knowledge." *Social Text* 20 (3): 67–78.

Shohat, Ella. 2004. "The 'Postcolonial' in Translation: Reading Said in Hebrew." *Journal of Palestine Studies* 33 (3): 55–75.

Shoshan, Malkit. 2010. *Atlas of the Conflict: Israel–Palestine* . Rotterdam: 010 Publishers.

Silverstein, Richard. 2010. "Oopsie, NGO Monitor Hauled into Court, Apologizes for Smearing Palestinian NGO." *Tikun Olam,* March 1.

Sismondo, Sergio. 2010. *An Introduction to Science and Technology Studies.* Chichester, UK: Wiley-Blackwell.

Smith, Andrea. 2013. "Unsettling the Privilege of Self-Reflexivity." In *Geographies of Privilege,* ed. France Winddance Twine and Bradley Gardener, 263–280. New York: Routledge.

Smith, Gregory W. H. 2006. *Erving Goffman.* Abingdon, UK: Routledge.

Smith, Merritt Roe, and Leo Marx, eds. 1984. *Does Technology Drive History? The Dilemma of Technological Determinism.* Cambridge, MA: MIT Press.

Smith, Neil. 1992. "Geography, Difference, and the Politics of Scale." In *Postmodernism and the Social Sciences,* ed. Joe Doherty, Elspeth Graham, and Mo Malek, 57–79. London: Macmillan.

Smith, Neil. 2008. *Uneven Development: Nature, Capital, and the Production of Space.* Athens: University of Georgia Press.

Söderström, Ola, Shalini Randeria, Didier Ruedin, Gianni D'Amato, and Francesco Panese, eds. 2013. *Critical Mobilities.* Lausanne: EPFL Press.

SOI (Survey of Israel). 1964. "Tel Aviv: Survey of Israel (SOI)." Drawer 88, Sheet 11-II. Jerusalem: Tel Aviv University Map Library.

SOI. 1970. *Atlas of Israel: Cartography, Physical Geography, Human and Economic Geography, History.* Jerusalem: SOI.

SOI. 2009. *Atlas Israel ha-Hadash: Ha-Atlas ha-Leumi* [The New Atlas of Israel: The National Atlas]. Jerusalem: SOI.

Soja, Edward W. 1996. *Thirdspace: Journeys to Los Angeles and Other Real-and-Imagined Places.* Oxford: Blackwell.

Soja, Edward W. 2000. *Postmetropolis: Critical Studies of Cities and Regions.* Oxford: Blackwell.

SOP (Survey of Palestine). 1945. *Maps and Publications.* Jaffa: SOP. Uncataloged. Tel Aviv University Map Library.

SOP. 1947. "Motor Map." Jaffa: SOP. 1:500,000. With Haganah Forces, Printed Maps and Photographs Services annotation of August 1948. Tel Aviv University Map Library.

SOP. 1948. *Report for the Years 1940–1946.* Jerusalem: SOP. Uncataloged. Tel Aviv University Map Library.

SOP. 1949. "Palestine: North Sheet" [annotated with 1949 Armistice Lines]. British Mandate Base Survey Drawing and Photo Process Office. Produced April 1946 from January 1946. SOP original. https://unispal.un.org/DPA/DPR/unispal.nsf/ 9a798adbf322aff38525617b006d88d7/f03d55e48f77ab698525643b00608d3 4/$FILE/Arm_1949.jpg (accessed March 3, 2016).

Sorkin, Michael. 2002. *The Next Jerusalem: Sharing the Divided City.* New York: Monacelli Press.

Spinner, Jackie. 2012. "The Cool New Palestinians: Geeks." *Christian Science Monitor,* February 18.

Spivak, Gayatri Chakravorty. 1988. "Can the Subaltern Speak?" In *Marxism and the Interpretation of Culture,* ed. Cary Nelson and Lawrence Grossberg, 271–316. Urbana: University of Illinois Press.

Spivak, Gayatri Chakravorty. 1999. *A Critique of Postcolonial Reason: Toward a History of the Vanishing Present.* Cambridge, MA: Harvard University Press.

Srebro, Haim. 2009. "A Status Report of the Activity of the Survey of Israel." Paper presented at the International Federation of Surveyors Working Week: Surveyors' Key Role in Accelerated Development, Eilat, May 3–8.

Srebro, Haim, Ron Adler, and Dov Gavish. 2009. *60 Years of Surveying and Mapping Israel: 1948–2008*. Tel Aviv: Survey of Israel.

St. Martin, Kevin, and John Wing. 2007. "The Discourse and Discipline of GIS." *Cartographica* 42 (3): 235–248.

Stamatopoulou-Robbins, Sophia. 2011. "In Colonial Shoes: Notes on the Material Afterlife in Post-Oslo Palestine." *Jerusalem Quarterly* 48:54–77.

Stein, Rebecca Luna. 1998. "National Itineraries, Itinerant Nations: Israeli Tourism and Palestinian Cultural Production." *Social Text* 56:91–124.

Stier, Michael. 1997. "Project GIS: The Planning and Design of a Geographical Information System within the 1995 Census." Paper presented at EUROSTAT and the Israeli Central Bureau of Statistics workshop, New Technologies for the 2000 Census Round: Sharing the Israeli Experience from the 1995 Census, Kibbutz Ma'ale Hachamisha, Israel, March 16.

Stoler, Ann Laura. 1995. *Race and the Education of Desire: Foucault's History of Sexuality and the Colonial Order of Things*. Durham, NC: Duke University Press.

Stoler, Ann Laura. 2010. *Along the Archival Grain: Epistemic Anxieties and Colonial Common Sense*. Princeton, NJ: Princeton University Press.

Stone, Jeffrey C. 1988. "Imperialism, Colonialism, and Cartography." *Transactions of the Institute of British Geographers* 13 (1): 57–64.

"Student's Journey from Gaza to Oxford." 2013. BBC, April 4.

Sufian, Sandra M. 2007. *Healing the Land and the Nation: Malaria and the Zionist Project in Palestine, 1920–1947*. Chicago: University of Chicago Press.

Sukarieh, Mayssoun, and Stuart Tannock. 2012. "On the Problem of Over-Researched Communities: The Case of the Shatila Palestinian Refugee Camp in Lebanon." *Sociology* 47 (3): 494–508.

Swedenburg, Ted. 1995. *Memories of Revolt: The 1936–39 Rebellion and the Palestinian National Past*. Minneapolis: University of Minnesota Press.

Swyngedouw, Erik. 2004. "Scaled Geographies: Nature, Place, and the Politics of Scale." In *Scale and Geographic Inquiry: Nature, Society, and Method*, ed. Eric Sheppard and Robert B. McMaster, 129–153. Oxford: Wiley-Blackwell.

Swyngedouw, Erik. 2006. "Metabolic Urbanism: The Making of Cyborg Cities." In *in The Nature of Cities: Urban Political Ecology and the Politics of Urban Metabolism*, ed. Nik Heynen, Maria Kaika, and Erik Swyngedouw, 21–40. New York: Routledge.

Szepesi, Stefan. 2012. *Walking Palestine: 25 Journeys into the West Bank*. Northampton, MA: Interlink Books.

Tamari, Salim. 2001. "Jerusalem: Subordination and Governance in a Sacred Geography." In *Capital Cities: Ethnographies of Urban Governance in the Middle East*, ed. Steney Shami, 175–203. Toronto: Centre for Urban and Community Studies, University of Toronto.

Tamari, Salim. 2009. "The Great War and the Erasure of Palestine's Ottoman Past." In *Transformed Landscapes: Essays on Palestine and the Middle East in Honor of Walid Khalidi*, ed. Camille Mansour and Leila Fawaz, 105–136. Cairo: American University in Cairo Press.

Tamari, Salim, and Elia Zureik. 2001. *Reinterpreting the Historical Record: The Uses of Palestinian Refugee Archives for Social Science Research and Policy Analysis*. Jerusalem: Institute of Jerusalem Studies.

Taraki, Lisa. 2008a. "Enclave Micropolis: The Paradoxical Case of Ramallah/al-Bireh." *Journal of Palestine Studies* 37 (4): 6–20.

Taraki, Lisa. 2008b. "Urban Modernity on the Periphery: A New Middle Class Reinvents the Palestinian City." *Social Text* 26 (2): 61–81.

Tawil-Souri, Helga. 2006. "Marginalizing Palestinian Development: Lessons against Peace." *Development* 49 (2): 75–80.

Tawil-Souri, Helga. 2011a. "Hacking Palestine: A Digital Occupation." *Al Jazeera English*, November 9.

Tawil-Souri, Helga. 2011b. "The Hi-Tech Enclosure of Gaza." Social Science Research Network Scholarly Paper 1764251, Rochester, NY.

Tawil-Souri, Helga. 2011c. "Orange, Green, and Blue: Color-Coded Paperwork for Palestinian Population Control." In *Surveillance and Control in Israel/Palestine: Population, Territory, and Power*, ed. Elia Zureik, David Lyon, and Yasmeen Abu-Laban, 219–238. London: Routledge.

Tawil-Souri, Helga. 2012. "Uneven Borders, Coloured (Im)mobilities: ID Cards in Palestine/Israel." *Geopolitics* 17:153–176.

Tawil-Souri, Helga, and Miriyam Aouragh. 2014. "Intifada 3.0? Cyber Colonialism and Palestinian Resistance." *Arab Studies Journal* 22 (1): 102–133.

Taylor, Peter J. 1990. "Editorial Comment GKS." *Political Geography Quarterly* 9 (3): 211–212.

Tesdell, Omar Imseeh. 2015. "Territoriality and the Technics of Drylands Science in Palestine and North America." *International Journal of Middle East Studies* 47 (3): 570–573.

Tesli, Arne. 2008. *Physical Planning and Institution Building: Lessons Learned and the Documentation of the PPIB Project in Palestine*. Oslo: Norwegian Institute for Urban and Regional Research.

Thomson, Janice E. 1996. *Mercenaries, Pirates, and Sovereigns: State-Building and Extraterritorial Violence in Early Modern Europe*. Princeton, NJ: Princeton University Press.

Todd, Jan. 1995. *Colonial Technology: Science and the Transfer of Innovation to Australia*. Cambridge: Cambridge University Press.

Trouillot, Michel-Rolph. 1994. *Silencing the Past: Power and the Production of History*. Boston: Beacon Press.

Tufakji, Khalil. n.d. "Maps and Survey Department." Orient House. http://www.orienthouse.org/dept/maps_dept.html (accessed October 1, 2015).

"Turkish Satellite to Roll Back Israel's Turf Veil." 2011. *Jerusalem Post* (Reuters), March 10.

Turnbull, David. 1996. "Cartography and Science in Early Modern Europe: Mapping the Construction of Knowledge Spaces." *Imago Mundi* 48:5–24.

Turnbull, David. 2000. *Masons, Tricksters, and Cartographers: Comparative Studies in the Sociology of Scientific and Indigenous Knowledge*. Amsterdam: Harwood.

United Nations Children's Emergency Fund. 2008. "Barrier Crossing Daunting for Deaf Palestinian Girl." Jerusalem, October 29.

UNOCHA (United Nations Office for the Coordination of Humanitarian Affairs). 2007. *The Humanitarian Impact on Palestinians of Israeli Settlements and Other Infrastructure in the West Bank*. Jerusalem: UNOCHA-OPT.

UNOCHA. 2008. *West Bank: Access and Closure*. Jerusalem: UNOCHA-OPT.

UNOCHA. 2009. *Shrinking Space: Urban Contraction and Rural Fragmentation in the Bethlehem Governorate*. Jerusalem: UNOCHA-OPT.

UNOCHA. 2010a. *Area C Humanitarian Response Plan Fact Sheet*. Jerusalem: UNOCHA-OPT.

UNOCHA. 2010b. *The Occupied Palestinian Territory: Overview Map*. Jerusalem: UNOCHA-OPT.

UNOCHA. 2011a. *East Jerusalem: Key Humanitarian Concerns*. Jerusalem: UNOCHA-OPT.

UNOCHA. 2011b. *West Bank: Area C Map*. Jerusalem: UNOCHA-OPT.

UNOCHA. 2012a. *Humanitarian Atlas*. Jerusalem: UNOCHA-OPT.

UNOCHA. 2012b. *West Bank Movement and Access Update*. Jerusalem: UNOCHA-OPT.

Urry, John. 2002. "Mobility and Proximity." *Sociology* 36 (2): 255–274.

US Department of State. 2010. "Knesset Considers Controversial NGO Legislation to Register as Foreign Agents." US Diplomatic cable from the Tel Aviv Embassy to the Office of the Secretary of State, Washington, DC. #10TELAVIV439. Wikileaks.

van Doorn, Niels, Sally Wyatt, and Liesbet van Zoonen. 2008. "A Body of Text." *Feminist Media Studies* 8 (4): 357–374.

Van Heur, Bas. 2009. "The Clustering of Creative Networks: Between Myth and Reality." *Urban Studies* 46 (8): 1531–1552.

Van Heur, Bas. 2010. *Creative Networks and the City: Towards a Cultural Political Economy of Aesthetic Production.* Bielefeld: Transcript.

Vannini, Phillip. 2011. "Constellations of Ferry (Im)mobility: Islandness as the Performance and Politics of Insulation and Isolation." *Cultural Geographies* 18 (2): 249–271.

van Teeffelen, Toine, and Fuad Giacaman. 2008. "Sumud: Resistance in Daily Life." In *Challenging the Wall: Toward a Pedagogy of Hope*, ed. Toine van Teeffelen. Bethlehem: Arab Educational Institute.

Vergano, Dan. 2013. "Contrary to Reports, Mount McKinley Not Shrinking." *National Geographic: Daily News*, November 13.

Visvanathan, Shiv. 1988. "Atomic Physics: The Career of an Imagination." In *Science, Hegemony, and Violence: A Requiem for Modernity*, ed. Ashis Nandy, 113–166. Delhi: Oxford University Press.

Wagner, Peter. 2001. *A History and Theory of the Social Sciences: Not All That Is Solid Melts into Air.* London: Sage.

Wallace, Timothy R., and Charles van den Heuvel. 2005. "Truth and Accountability in Geographic and Historical Visualization." *Cartographic Journal* 42 (2): 173–181.

Warf, Barney, and Santa Arias, eds. 2009. *The Spatial Turn: Interdisciplinary Perspectives.* London: Routledge.

Watzman, Haim. 1993. "Israel Floods Drained Swamp to Bring in Tourists." *New Scientist*, April 17.

Webb, Eugene J., Donald T. Campbell, Richard D. Schwartz, and Lee Sechrest. 1966. *Unobtrusive Measures.* Chicago: Rand McNally.

Wegman, Edward J., and Daniel B. Carr. 1993. "Statistical Graphics and Visualization." In *Handbook of Statistics 9: Computational Statistics*, ed. C. N. R. Rao, 857–958. Amsterdam: Elsevier.

Weizman, Eyal. 2002. "The Politics of Verticality." https://www.opendemocracy.net/conflict-politicsverticality/debate.jsp (accessed October 28, 2015).

Weizman, Eyal. 2007. *Hollow Land: Israel's Architecture of Occupation*. London: Verso.

Weizman, Eyal. 2011. *The Least of All Possible Evils: Humanitarian Violence from Arendt to Gaza*. London: Verso.

Weizman, Eyal. 2014. *Forensis: The Architecture of Public Truth*. Berlin: Sternberg Press.

Wild Olive. n.d. "Who's [sic] Land Is It Anyway?" http://www.wildolive.co.uk/land.htm (accessed October 21, 2015).

Winichakul, Thongchai. 1994. *Siam Mapped: A History of the Geo-Body of a Nation*. Honolulu: University of Hawaii Press.

Winner, Langdon. 1980. "Do Artifacts Have Politics?" *Daedalus* 109 (1): 121–136.

Wittgenstein, Ludwig. 2001. *Philosophical Investigations*. Oxford: Blackwell.

Wolf, Aaron, and John Ross. 1992. "The Impact of Scarce Water Resources on the Arab–Israeli Conflict." *Natural Resources Journal* 32:919–958.

Wolfson, Zev. 2005. "Funders Ignoring Deceit by Palestinian Eco-NGOs." *NGO Monitor Digest* 3 (12): n.p.

Wood, Denis. 2010. *Rethinking the Power of Maps*. New York: Guilford Press.

Wyatt, Sally. 1998. "Technology's Arrow: Developing Information Networks for Public Administration in Britain and the United States." PhD diss., Maastricht University.

Wyatt, Sally. 2003. "Non-Users Also Matter: The Construction of Users and Non-Users of the Internet." In *How Users Matter: The Co-Construction of Users and Technology*, ed. Nelly Oudshoorn and Trevor Pinch, 67–79. Cambridge, MA: MIT Press.

Wyatt, Sally. 2008a. "Challenging the Digital Imperative." Inaugural lecture presented on the acceptance of the Royal Netherlands Academy of Arts and Sciences Extraordinary Chair in Digital Cultures in Development, Maastricht University, March 28.

Wyatt, Sally. 2008b. "Technological Determinism Is Dead; Long Live Technological Determinism." In *Handbook of Science and Technology Studies*, ed. Edward J. Hackett, Olga Amsterdamska, Michael E. Lynch and Judy Wajcman, 165–180. Cambridge, MA: MIT Press.

Wyatt, Sally, and Brian Balmer. 2007. "Home on the Range: What and Where Is the Middle in Science and Technology Studies?" *Science, Technology, and Human Values* 32 (6): 619–626.

Wyatt, Sally, Andrea Scharnhorst, Anne Beaulieu, and Paul Wouters. 2013. "Introduction to Virtual Knowledge". In *Virtual Knowledge: Experimenting in the Humanities and the Social Sciences*, ed. Paul Wouters, Anne Beaulieu, Andrea Scharnhorst, and Sally Wyatt, 1–23. Cambridge, MA: MIT Press.

Wyly, Elvin. 2011. "Positively Radical." *International Journal of Urban and Regional Research* 35 (5): 889–912.

Yakin, Boaz, and Nick Bertozzi. 2013. *Jerusalem: A Family Portrait.* New York: First Second.

Yaron, Oded. 2011. "Israel Allows Google to Operate Controversial Street View." *Ha'aretz*, August 21.

Yaron, Oded. 2012. "Google to Launch Street View for Israel, with Government Nod." *Ha'aretz*, April 16.

Yousef, Omar. 2011. "From Salah Eddin to A-Ram, an Everyday Journey." *Palestine–Israel Journal of Politics, Economics, and Culture* 17 (12): n.p.

Yuval-Davis, Nira. 2012. "Dialogical Epistemology: An Intersectional Resistance to the 'Oppression Olympics.'" *Gender and Society* 26 (1): 46–54.

Zawawi, Zahraa, Eric Corijn, and Bas Van Heur. 2013. "Public Spaces in the Occupied Palestinian Territories." *GeoJournal* 78 (4): 743–758.

Zonszein, Mairav. 2013a. "Official Admits East Jerusalem Park Meant to Stop Palestinian Expansion." *+972 Magazine*, September 30.

Zonszein, Mairav. 2013b. "Watch: Israeli Activists Detained for Filming Illegal Settlement Construction." *+972 Magazine*, April 3.

Zrahiya, Zvi. 2011. "How Can Israelis Protect Their Privacy with Google Street View on the Prowl?" *Ha'aretz*, February 22.

Zukin, Sharon. 1991. *Landscapes of Power: From Detroit to Disney World.* Berkeley: University of California Press.

Zureik, Elia. 2001. "Constructing Palestine through Surveillance Practices." *British Journal of Middle Eastern Studies* 28 (2): 205–227.

Zureik, Elia, David Lyon, and Yasmeen Abu-Laban, eds. 2011. *Surveillance and Control in Israel/Palestine: Population, Territory, and Power.* London: Routledge.

Index

Note: Page numbers in *italics* indicate illustrations.

Inside Technology

Edited by Wiebe E. Bijker, W. Bernard Carlson, and Trevor Pinch

Cyrus C. M. Mody, *Instrumental Community: Probe Microscopy and the Path to Nanotechnology*

Morana Alač, *Handling Digital Brains: A Laboratory Study of Multimodal Semiotic Interaction in the Age of Computers*

Gabrielle Hecht, editor, *Entangled Geographies: Empire and Technopolitics in the Global Cold War*

Michael E. Gorman, editor, *Trading Zones and Interactional Expertise: Creating New Kinds of Collaboration*

Matthias Gross, *Ignorance and Surprise: Science, Society, and Ecological Design*

Andrew Feenberg, *Between Reason and Experience: Essays in Technology and Modernity*

Wiebe E. Bijker, Roland Bal, and Ruud Hendricks, *The Paradox of Scientific Authority: The Role of Scientific Advice in Democracies*

Park Doing, *Velvet Revolution at the Synchrotron: Biology, Physics, and Change in Science*

Gabrielle Hecht, *The Radiance of France: Nuclear Power and National Identity after World War II*

Richard Rottenburg, *Far-Fetched Facts: A Parable of Development Aid*

Michel Callon, Pierre Lascoumes, and Yannick Barthe, *Acting in an Uncertain World: An Essay on Technical Democracy*

Ruth Oldenziel and Karin Zachmann, editors, *Cold War Kitchen: Americanization, Technology, and European Users*

Deborah G. Johnson and Jameson W. Wetmore, editors, *Technology and Society: Building Our Sociotechnical Future*

Trevor Pinch and Richard Swedberg, editors, *Living in a Material World: Economic Sociology Meets Science and Technology Studies*

Christopher R. Henke, *Cultivating Science, Harvesting Power: Science and Industrial Agriculture in California*

Helga Nowotny, *Insatiable Curiosity: Innovation in a Fragile Future*

Karin Bijsterveld, *Mechanical Sound: Technology, Culture, and Public Problems of Noise in the Twentieth Century*

Peter D. Norton, *Fighting Traffic: The Dawn of the Motor Age in the American City*

Joshua M. Greenberg, *From Betamax to Blockbuster: Video Stores and the Invention of Movies on Video*

Mikael Hård and Thomas J. Misa, editors, *Urban Machinery: Inside Modern European Cities*

Christine Hine, *Systematics as Cyberscience: Computers, Change, and Continuity in Science*

Wesley Shrum, Joel Genuth, and Ivan Chompalov, *Structures of Scientific Collaboration*

Shobita Parthasarathy, *Building Genetic Medicine: Breast Cancer, Technology, and the Comparative Politics of Health Care*

Kristen Haring, *Ham Radio's Technical Culture*

Atsushi Akera, *Calculating a Natural World: Scientists, Engineers, and Computers during the Rise of US Cold War Research*

Donald MacKenzie, *An Engine, Not a Camera: How Financial Models Shape Markets*

Geoffrey C. Bowker, *Memory Practices in the Sciences*

Stuart Blume, *Insight and Industry: On the Dynamics of Technological Change in Medicine*

Donald MacKenzie, *Inventing Accuracy: A Historical Sociology of Nuclear Missile Guidance*

Pamela E. Mack, *Viewing the Earth: The Social Construction of the Landsat Satellite System*

H. M. Collins, *Artificial Experts: Social Knowledge and Intelligent Machines*

http://mitpress.mit.edu/books/series/inside-technology